Wind Load Design for Petrochemical and Other Industrial Facilities

Second Edition

Prepared by
Task Committee on Wind-Induced
Forces of the Oil and Gas Committee of the
Energy Division of the American Society of Civil Engineers

ASCE

Published by the American Society of Civil Engineers

Library of Congress Cataloging-in-Publication Data

Names: American Society of Civil Engineers, author.
Title: Wind load design for petrochemical and other industrial facilities / Task Committee on Wind-Induced Forces of the Oil and Gas Committee of the Energy Division of the American Society of Civil Engineers.
Other titles: Wind loads for petrochemical and other industrial facilities
Description: Second edition. | Reston, Virginia : American Society of Civil Engineers, [2020] | Revised edition of: Wind loads for petrochemical and other industrial facilities / prepared by Task Committee on Wind-Induced Forces of the Petrochemical Committee of the Energy Division of the American Society of Civil Engineers, 2011. | Includes bibliographical references and index. | Summary: "This report represents the state-of-the-practice for wind load design at industrial facilities and is aimed at engineers familiar with design of industrial type structures"– Provided by publisher.
Identifiers: LCCN 2020012756 | ISBN 9780784415610 (paperback) | ISBN 9780784483039 (adobe pdf)
Subjects: LCSH: Petroleum refineries–Design and construction. | Industrial buildings–Design and construction. | Petroleum refineries–Aerodynamics. | Industrial buildings–Aerodynamics. | Wind-pressure.
Classification: LCC TH4571 .A44 2020 | DDC 624.1/75–dc23
LC record available at https://lccn.loc.gov/2020012756

Published by American Society of Civil Engineers
1801 Alexander Bell Drive
Reston, Virginia 20191-4382
www.asce.org/bookstore | ascelibrary.org

Errata: Errata, if any, can be found at https://doi.org/10.1061/9780784415610.

ISBN 978-0-7844-1561-0 (print)
ISBN 978-0-7844-8303-9 (PDF)

Manufactured in the United States of America.

26 25 24 23 22 21 20 1 2 3 4 5

ASCE Oil and Gas Committee

This publication began as one of five state-of-the-practice engineering reports produced by the former ASCE Petrochemical Energy Committee. These engineering reports were intended to be a summary of current engineering knowledge and design practice, and present guidelines for the design of petrochemical facilities. They represented a consensus opinion of task committee members active in their development. These five ASCE engineering reports were initially published in 1997 in four separate books, with the reports on wind and anchor bolts being printed together in a combined document for publishing convenience, they are:

1. *Design of Anchor Bolts in Petrochemical Facilities*
2. *Design of Blast Resistant Buildings in Petrochemical Facilities*
3. *Design of Secondary Containment in Petrochemical Facilities*
4. *Guidelines for Seismic Evaluation and Design of Petrochemical Facilities*
5. *Wind Loads on Petrochemical Facilities*

Since the initial publication of these five reports, buildings codes and standards have changed significantly, specifically in the calculation of wind and seismic loads and analysis procedures for anchorage design. Additionally, new research in these areas and in blast-resistant design provided opportunities for improvement of the recommended guidelines. ASCE then determined the need to update four of the original reports and publish new editions, based on the latest research and for consistency with current building codes and standards. The ASCE Petrochemical Energy Committee was hence reorganized in 2005 and the following four task committees were formed to update their respective reports:

- Task Committee on Anchorage Design
- Task Committee on Blast-Resistant Design
- Task Committee on Seismic Evaluation and Design for Petrochemical Facilities
- Task Committee on Wind-Induced Forces

In 2011, ASCE published a subsequent version of the renamed *Wind Loads for Petrochemical and Other Industrial Facilities*. However, because the 1997 edition of the book was published in a combined document with *Anchorage Design for Petrochemical Facilities* and the 2011 version of the book was a standalone document, the 2011 version was officially called a first edition (of the standalone document).

Since the publication of that report, notable updates to commonly used design codes and standards cited in the book have been issued. As a result, in 2016 the ASCE Energy Division Executive Committee directed that a task committee be formed to update the book to ensure consistency with current building codes and standards.

In January 2017, James R. (Bob) Bailey, chair of the Oil and Gas Committee acting on behalf of the Energy Division Executive Committee, reformed the Task Committee on Wind-Induced Forces to update the book, resulting in the issuance of this present book, *Wind Load Design for Petrochemical and Other Industrial Facilities*, Second Edition. Although titled a second edition (of the standalone document), it is the third time that the book has been issued (1997, 2011, and now 2020).

ASCE Energy Division Executive Committee

David Kerins, *Chair*	ExxonMobil
Hongchun Liu, *Vice Chair*	Bechtel
Greg J. Soules, *Past Chair (expired)*	CB&I
James R. (Bob) Bailey	Exponent, Inc.
Jason Hedien	Stantec Consulting Services
Dennis Truax	Mississippi State University
Silky S. K. Wong	Dow (formerly with Fluor Corp.)

ASCE Task Committee on Wind-Induced Forces

This book is intended to be a state-of-the-practice set of guidelines. It is based on reviews of current practice, internal company standards, published documents, and the work of related organizations. The book includes references that provide additional information, commercial publications, and government reports.

This book was prepared to provide guidance for determination of wind-induced forces on structures found in petrochemical and other industrial facilities. It should be of interest to engineers familiar with the design of industrial-type structures and the application of ASCE 7-16, *Minimum Design Loads and Associated Criteria for Buildings and Other Structures*, to these types of structures.

In helping create a consensus set of guidelines, a number of individuals provided valuable assistance and review. Reviewers of the entire manuscript include John Geigel, P.E. (retired, formerly ExxonMobil) and Clay Willis, P.E., S.E. (Wood Group USA, Inc.). Reviewers of the section titled "Wind Loads for LNG Facilities," also include Joseph Sieve [Pipeline and Hazardous Materials Safety Administration (PHMSA), US Department of Transportation] and Robert Bachman, P.E., S.E. (R.E. Bachman, Consulting Structural Engineer). The committee is extremely appreciative of the efforts of these reviewers.

The committee is also very appreciative of the efforts of its representative on the main Oil and Gas Committee, Bob Bailey, Ph.D., P.E., F.ASCE, who contributed knowledge and guidance to the committee and attended committee meetings.

Finally, the committee would also like to acknowledge Jay Snyder and Donna Dickert of ASCE, who provided logistical facilitation along the way.

Paul B. Summers, P.E., S.E., F.ASCE, Simpson Gumpertz & Heger, *Chair*
Samuel D. Amoroso, Ph.D., P.E., S.E., Exponent, Inc., *Co-Chair*
Matthew J. Brightman, P.E., S.E., ExxonMobil, *Secretary*

Guzhao Li, Ph.D., P.E., S.E.	Simpson Gumpertz & Heger
Rudy Mulia, P.E., S.E.	Chevron Energy Technology Company
Carlos A. Ortega, Ph.D., P.E., S.E.	Simpson Gumpertz & Heger
Rajendra Prasad, P.E.	Wood Group USA, Inc
Nicholas Truong, P.E.	Cheniere Energy, Inc.
Silky S. K. Wong, Ph.D., P.E., S.E.	Dow (formerly with Fluor Corp.)
Randall L. Wright, P.E., S.E.	Wood Group USA, Inc.
Xiping Steven Wu, Ph.D.	Shell Projects & Technology

Contents

Chapter 1 Introduction **1**
1.1 Background 1
1.2 State of the Practice 3
1.3 Purpose of This Book 3
References 3

Chapter 2 Background **5**
2.1 Introduction 5
2.2 Key Wind Engineering Concepts 5
2.3 Aerodynamics of Open-Frame Structures 11
2.4 Aerodynamics of Partially Clad Structures 13
2.5 Aerodynamics of Vertical Vessels 14
2.6 Other Wind-Loading Codes, Standards, and Guides 17
2.7 Research Progress and Future Needs 18
References 20

Chapter 3 Review of Existing Design Practices **21**
3.1 Introduction 21
3.2 Survey of Existing Practices and Impacts of First Edition of Guidelines 21
3.3 Existing Design Practices for Structures Not Covered by This Guide 22
References 30

Chapter 4 Recommended Guidelines Part I: Design Considerations and Methods **33**
4.1 Introduction 33
4.2 Historical Performance 33
4.3 Wind Tunnel Testing 44
4.4 CFD Commentary 47
4.5 Load Combinations 49
4.6 Wind Loads for LNG Facilities 62
4.7 Evaluation of Wind Loads on Existing Structures 69
4.8 Wind Load Analysis Uncertainty 75
References 75

Chapter 5 Recommended Guidelines Part II: Analytical Determination of Wind Loads 77
5.1 General 77
5.2 Pipe Racks 78
5.3 Open-Frame Structures 80
5.4 Partially Clad Structures 94
5.5 Pressure Vessels 96
5.6 Cooling Towers 106
5.7 Air-Cooled Heat Exchanger (Air Coolers or Fin-Fans) 107
5.8 Compressor Shelters 109
Appendix 5A. Alternate Method for Determining C_f and Load Combinations for Open-Frame Structures 110
Appendix 5B. High-Solidity Open-Frame Structures 117
References 120

Chapter 6 Examples 123
6.1 Introduction 123
6.2 Pipe Rack and Pipe Bridge Example 124
6.3 Open-Frame Examples 133
6.4 Partially Clad Structure Example 148
6.5 Pressure Vessels Example 149
6.6 Cooling Tower Example 161
6.7 Air Cooler (Fin-Fan) Example 165
References 167

Index 169

CHAPTER 1
Introduction

The focus of this book is on the procedures for determining the design wind loads for non-building structures in petrochemical and other industrial facilities. The book is structured around the following generic types of structures usually found in these facilities. Examples are also provided for some of these structures:

- Pipe support structures (pipe racks, pipe bridges),
- Open and partially clad frame structures,
- Vessels (vertical, horizontal, and spherical),
- Cooling towers,
- Air coolers (air-cooled heat exchangers, also known as fin-fans),
- Tanks, and
- Steel stacks.

1.1 BACKGROUND

The basis and procedures for determining design wind loads for enclosed structures and other conventional structures are well documented in the engineering literature. These design bases and procedures have been adopted by ASCE and prescribed in ASCE 7-16, *Minimum Design Loads and Associated Criteria for Buildings and Other Structures* (hereafter referred to as ASCE 7) and its predecessor documents. Other organizations have incorporated the major provisions of ASCE 7 into building codes. The International Building Code (IBC) states that wind loads should be calculated in accordance with the latest version of ASCE 7, and the IBC is adopted throughout the United States. ASCE 7 provides several methods for calculating design wind loads on the main wind force resisting system (MWFRS) and on components and cladding:

- Directional procedure for building MWFRS;
- Envelope procedure for building MWFRS;
- Directional procedure for MWFRS of building appurtenances and other structures;

- Standard methods for component and cladding loads for low-rise buildings, buildings taller than 60 ft, open buildings, building appurtenances, rooftop structures, and equipment;
- Simplified method for component and cladding loads for low-rise buildings; and
- Wind tunnel procedure.

The envelope procedure for building MWFRS (ASCE 7, Chapter 28) was introduced to simplify the analysis of typical regular-shaped building structures. Its use is restricted to relatively rigid, low-rise, enclosed structures. The directional procedure for building MWFRS (ASCE 7, Chapter 27) is permitted for structures of any height that do not have unusual geometric irregularities or unusual response characteristics. The wind tunnel procedure may be required for complex structures that cannot be evaluated using the other five methods in ASCE 7.

The scope statement in ASCE 7, Section 1.1 indicates that the standard provides minimum load requirements for the design of buildings, other structures, and their nonstructural components that are subject to building codes. The standard also addresses enclosed structures, trussed towers, and simple cylinders commonly found in petrochemical facilities. However, to address important nonbuilding structures in petrochemical and other industrial facilities, this book enhances ASCE 7-16 provisions for open-frame structures, structures with interconnecting piping, partially clad structures, vessels with attached piping and platforms, cooling towers, and air coolers. Design wind loads on nonbuilding structures are typically calculated using the force equation from ASCE 7:

$$F = q_z \, G \, C_f \, A_f \qquad (29.4\text{-}1)$$

where

q_z = Velocity pressure component,
G = Gust effect factor,
C_f = Force/shape/drag/shielding component, and
A_f = Area for which the force is calculated that is usually the projected area normal to the wind.

The velocity pressure, q_z, is calculated using ASCE 7-16 based on several factors, such as the risk category of the structure, the surrounding terrain (exposure category), and the basic wind speed, among others.

The selection of risk category, basic wind speed, exposure category, gust effect factor, and other factors is described in ASCE 7; therefore, it is not discussed in detail herein. This book expands on ASCE 7's coverage of force coefficients, projected areas, and shielding for industrial-type structures and equipment, which must be carefully defined to ensure that behavior under wind forces is taken into account. These wind load components are discussed within and recommendations for selecting values are made. Because this book is intended to supplement ASCE 7, the designer is referred to that standard when it provides the appropriate information. The nomenclature used in the recommendations herein mirrors that found in ASCE 7.

1.2 STATE OF THE PRACTICE

This book reflects various company practices, available research, and committee consensus for the wind load design of petrochemical and other industrial facilities. The committee responsible for *Wind Loads for Petrochemical and Other Industrial Facilities* (ASCE 2011) performed a survey (see Section 3.2), and the results indicate that more than half of the companies surveyed had design practices that referenced *Wind Loads and Anchor Bolt Design for Petrochemical Facilities*, the first version of this report (ASCE 1997). These survey results are evidence that there was increased uniformity in determining wind loads for petrochemical and other industrial facilities during the decades between the first and second versions of this book.

1.3 PURPOSE OF THIS BOOK

It is the intent of this committee that the publication of this book will continue to progress a more uniform application of practices for the computation of design wind loads for petrochemical and industrial facilities. To facilitate this goal, a set of recommended guidelines is presented as part of the book.

References

ASCE. 1997. *Wind loads and anchor bolt design for petrochemical facilities.* Reston, VA: ASCE.

ASCE. 2011. *Wind loads for petrochemical and other industrial facilities.* Reston, VA: ASCE.

ASCE. 2017. *Minimum design loads and associated criteria for buildings and other structures.* ASCE 7-16. Reston, VA: ASCE.

ICC (International Code Council). 2018. *International building code (IBC).* Washington, DC: ICC.

CHAPTER 2
Background

2.1 INTRODUCTION

This chapter provides background information that is intended to assist the user of this book in the interpretation and application of the material that is presented in subsequent chapters. The sections of this chapter include discussions of fundamental wind engineering concepts and the general aerodynamic characteristics of structure types common to petrochemical and other industrial facilities. The chapter concludes with a presentation of other sources of guidance that may be helpful to designers or analysts, a brief summary of research progress since the first edition, and research needs for the future.

2.2 KEY WIND ENGINEERING CONCEPTS

2.2.1 Wind Load Formulation

The wind load on a structure is a function of many different variables. The purpose of this section is to review the basic wind load formulation in ASCE 7 as it pertains to petrochemical structures, as well as briefly discuss some considerations relevant to each of the variables. Equation (29.4-1) in ASCE 7 is used to calculate the wind force on other, nonbuilding structures such as petrochemical structures. This equation is shown as follows.

$$F = q_z \cdot G \cdot C_f \cdot A_f \quad (29.4\text{-}1)$$

where

q_z = Velocity pressure,
G = Gust effect factor,
C_f = Aerodynamic force coefficient, and
A_f = Reference area.

The velocity pressure, q_z, as defined Section 26.10 of ASCE 7 and shown in the following equation in US customary units (for the metric equivalent of this equation, see ASCE 7-16) is a modification of Bernoulli's formula, which relates the fluid dynamic pressure to the fluid density, and the square of the flow velocity (or in this case, the basic wind speed). The constant at the beginning of the velocity pressure equation in ASCE 7 includes the density of air and a factor for the conversion of units such that when the wind speed is specified in units of miles per hour, the resulting velocity pressure is in units of pounds per square foot. The remaining variables in the velocity pressure equation are dimensionless.

$$q_z = 0.00256 \cdot K_z \cdot K_{zt} \cdot K_d \cdot K_e \cdot V^2 \qquad (26.10\text{-}1)$$

Of all the variables in this equation, the specification of the basic wind speed, V, has the greatest influence on the velocity pressure, q_z, and the resulting wind load, F. Basic wind speeds are specified by ASCE 7 as 3 s gust wind speeds at standard meteorological heights of 33 ft (10 m) in open terrain. The nominal return periods [or mean recurrence intervals (MRIs)] vary from 300 to 3,000 years, depending on the risk category of the structure. The inverse of the MRI for a given wind speed in years is the probability that such a wind speed will occur in any given year. Because some geographic locations are more susceptible to severe winds than others owing to variations in local climatology, the wind speed associated with a particular MRI in general will also vary with geographic location.

ASCE 7, Figures 26.5-1 and 26.5-2 are maps of the United States with contours of basic wind speed (isotachs) and political boundaries resolved to the state and county levels. This map facilitates easy selection of basic wind speeds for use in ASCE 7, Equation (26.10-1). Because of the general lack of reliable meteorological measurements in the hurricane-prone regions of the Atlantic and Gulf coasts of the United States, the basic wind speeds for these locations have been estimated through Monte Carlo simulation using mathematical models of hurricane wind fields and statistical data gathered over the past 150 or so years. Special wind regions that have unique meteorological influences are identified on the map in ASCE 7, Figures 26.5-1 and 26.5-2. Specification of the basic wind speed for special wind regions requires consultation with local building officials.

Wind speeds typically increase with height above the ground surface. This variation occurs because of the influence of friction and viscosity on the wind field. The rate at which the wind speed increases with height is governed primarily by the roughness of the terrain over which it is flowing. This roughness may be in the form of elements of the built environment, vegetation, or water waves on large bodies of water located in the upwind fetch. The combined effects of structure height and surface roughness are incorporated into the estimate of the velocity pressure, q_z, through the exposure factor, K_z. This factor is specified in ASCE 7 for three different terrain exposures and for a variety of heights. The three terrain exposures correspond to urban/suburban or forested terrain (Exposure B), open terrain with scattered, low obstructions (Exposure C), and very flat terrain or shoreline locations (Exposure D). The value of the exposure factor increases with increasing height above the ground and decreasing surface roughness. Because the basic wind speed, V, is defined at the standard height of 33 ft (10 m) in open terrain, the value of K_z for Exposure C at a height of 33 ft is 1.0.

Topographic effects can influence the wind loading of structures by disturbing the flow field near features such as hills, ridges, and escarpments, resulting in local deviations from the basic wind speed. In general, these changes may increase or decrease the local wind speed, but the ASCE 7 wind load formulation conservatively only considers cases in which topographic features tend to increase local wind velocities, and thus wind loads. These effects are incorporated into the wind load formulation through an adjustment of the velocity pressure by the topographic effect factor, K_{zt}. ASCE 7 provides methods for calculating the value of this factor for some simple special cases. The engineer should consider whether or not these special cases adequately represent the conditions for the location of interest. For more complicated topography, the engineer may need to exercise careful engineering judgment, seek expert consultation, or resort to either physical or numerical modeling to understand the topographic influences on the wind loads.

The directionality factor, K_d, accounts for the lower probability that the most vulnerable orientation for a structure will correspond unfavorably with the direction of maximum wind speed during a design event. As such, the value of K_d is always less than 1.0, but varies with structure type. ASCE 7 provides tabulated values of the directionality factor for a variety of structural forms (Table 26.6-1). The values for K_d are defined only for use with the load combinations listed in Section 2.3 and 2.4 of the standard when designing a structure. It is the opinion of this committee that the load combinations and examples presented in this book should use the relevant ASCE 7 values for K_d. The use of the directionality factor may not be warranted when using ASCE 7 methods in an analysis of the performance of an existing structure for which the wind direction is specified or known (e.g., an analysis in a forensic investigation). For some cases relevant to petrochemical and other industrial facilities, this committee has recommended directionality factor values for circumstances in addition to those covered in the standard.

The coefficient 0.00256 in ASCE 7, Equation (26.10-1) incorporates the density of air at sea level. Because the density of air decreases with ground elevation, users are permitted to apply a ground elevation factor, K_e, for structures situated at higher elevations. ASCE 7, Table 26.9-1 summarizes some values of K_e. It is permitted to conservatively take K_e as 1.0 for all elevations.

The gust effect factor, G, simultaneously incorporates two different processes through which the turbulence in the atmospheric wind can affect wind loads for a structure. The first of these processes is related to the size of a structure relative to significant scales that appear in the turbulence. By definition, a gust of wind is short-lived. Relating wind speed and length of time required for passage of a gust gives some indication of the relative dimensions of the gust. Of course, turbulence, or gustiness, is a chaotic process, and these characteristics can be defined only in a statistical sense. If a structure is large relative to the significant scales of the wind gusts, the building will often not be completely enveloped by a single gust of wind. As such, the maximum pressures that occur on different building surfaces will not be strictly correlated in time, and the total resulting wind load will be lower than what would be expected by considering the gust wind speed to apply to all the building surfaces simultaneously. ASCE 7 provides a method for calculating how the relationship between building dimensions and the turbulence characteristics of the wind field influence the gust effect factor.

In addition to the spatial considerations related to wind gusts, the relative flexibility of the structure combined with the temporal fluctuations in the flow velocity may cause certain structures to be dynamically excited in the along-wind direction. If the fundamental frequency of a structure is low enough to enter the range of significant frequency content associated with the wind turbulence, the structure can experience some load amplification owing to resonance. ASCE 7 also provides a method for calculating how the relationship between building fundamental frequency and the turbulence characteristics of the wind field influence the gust effect factor. These dynamic effects can be neglected if a structure is sufficiently rigid. ASCE 7 defines a rigid structure as one with a fundamental frequency greater than 1 Hz. The structures that may be susceptible to along-wind dynamic excitation in petrochemical and other industrial facilities are tall, slender structures such as vertical vessels, flare towers, stacks, and structures with heavy equipment located or supported near the top of the structure. The discussion on gust effects in Section 2.5 and the example calculation for flexible vertical vessels (Section 6.5.1.3 of this book) provide additional commentary regarding this factor.

The two remaining variables in the wind load formulation are the force coefficient, C_f, and the projected area, A_f. The force coefficient is an empirical factor that typically is determined through experiment for the structure or shape of interest. The selection of values for these variables as they relate to petrochemical structures is the focus of Chapter 5 of this book.

In addition to dependence on structure shape, the force coefficient, C_f, also depends on a nondimensional flow parameter called the Reynolds number. The Reynolds number (*Re*) is a ratio of the inertial forces in a flow field to the viscous forces in the flow field, and is defined as follows:

$$Re = \frac{U \cdot D}{\nu} \tag{2-1}$$

where

U = Flow velocity,
D = Characteristic dimension for the structure or element, and
ν = Kinematic viscosity.

In the range of *Re* that is typical of full-scale flows for structural applications, many structural forms will not exhibit much sensitivity of the force coefficient to *Re*. This phenomenon is particularly true for rectangular bodies or structures with sharp geometries. Structures with curved surfaces generally show more sensitivity of the force coefficient to *Re*. For example, the force coefficient for a cylindrical section comprising a handrail component may be substantially larger than the force coefficient for the body of a vertical vessel, which is also cylindrical in form. For the same wind environment, the difference in diameters may be two orders of magnitude, resulting in a proportional difference in *Re* for the flow near each of these elements. The flow fields and associated force coefficients for cylindrical bodies exhibit dramatic variability for a particular range of *Re* (10^5 to 10^6). As an example, the force coefficient data provided in Figure 29.4-1 of ASCE 7-16 accounts for these effects for structures such as round chimneys and tanks.

2.2.2 Wind Speed Interpretations

The design wind speeds used in the wind load formulation in ASCE 7 are specified as 3 s gust wind speeds. This nomenclature refers to the time over which wind speed measurements are averaged. For example, if an anemometer (an instrument for measuring wind speed) records data at a frequency of 10 Hz (i.e., 10 readings per second), then the 3 s gust wind speed would be the arithmetic mean of 30 consecutive individual measurements from that device. For fluctuating wind speed signals typical of atmospheric conditions, individual measurements would be higher or lower than the resulting mean value over the averaging interval. Furthermore, it will always be possible to find a subinterval (say, 1 s in this example) that has a mean wind speed higher than that of the interval itself.

This concept extends to other averaging times as well. "Sustained" wind speeds are conventionally defined on the basis of a 1 min averaging time. For any given sustained wind speed measurement, there will be a higher maximum 3 s gust wind speed in the 60 s interval corresponding to the measurement. Gust factors (not to be confused with the gust *effect* factors described previously) are used to convert wind speeds based on one averaging time to equivalent wind speeds corresponding to another averaging time. When converting wind speeds from longer to shorter averaging times, the gust factors will always be greater than 1.0. The commentary of ASCE 7 includes a figure that can be used to determine gust factors for converting between equivalent wind speeds for averaging times ranging from 1 s to 1 h. As an example, the factor for converting a 1 min "sustained" wind speed to an equivalent 3 s gust wind speed is approximately 1.22 for synoptic winds and 1.28 for hurricane winds according to Figure C26.5-1 in the ASCE 7 commentary.

A wide variety of wind speed averaging times are represented in engineering standards and through meteorological reporting. ASCE 7 specified design wind speeds on the basis of the "fastest mile of wind" prior to the 1995 edition of the standard. The fastest-mile specification actually has a variable averaging time, which is as follows:

$$t = \frac{3{,}600}{V_{fm}} \tag{2-2}$$

where t is the averaging time in seconds, and V_{fm} is the fastest-mile wind speed in miles per hour.

The averaging time is equal to 60 s for wind speeds of 60 mi/h but increases for lower wind speeds and decreases for higher wind speeds. Other international engineering standards have specified design wind speeds on the basis of a mean hourly value or a 10 min mean. When reviewing older designs or when analyzing structures at international facilities, it is important to recognize that these variations may be present.

Meteorological measurements located at US airports have historically reported gust and sustained wind speeds based on 5 s and 2 min averaging times, respectively. As technology has changed, mechanical cup anemometers are being replaced with sonic anemometers at these locations, and the reference averaging times for the measurements are being changed to 3 s and 1 min, respectively. Hurricane intensity is conventionally defined on the basis of the fastest 1 min mean wind speed occurring anywhere in the storm [at a reference height of 10 m (33 ft) above open water]. The commentary to ASCE 7 contains a table providing conversions of the sustained wind speeds for each of the Saffir–Simpson hurricane categories to their equivalent 3 s gust values.

2.2.3 Crosswind Forces

The ASCE 7 wind load formulation for other, nonbuilding structures is applicable to loadings along the nominal direction of the wind. The standard contains no analytical provisions for estimating wind loads that develop in a direction perpendicular to the wind direction for these types of structures. There are some common instances in which such forces may be significant. If a structure is asymmetric in plan with respect to the wind direction, it is likely that wind loads will be generated perpendicular to the flow. This scenario may exist for a single structure, or for a structure in a group that is oriented asymmetrically with respect to the wind direction. The aerodynamic action responsible for the development of these types of crosswind forces is similar to development of lift on the surface of an airplane wing. The unequal velocity and pressure distributions that develop on asymmetric bodies lead to net forces acting perpendicular to the flow direction. These across-wind forces act simultaneously with the along-wind drag forces.

Another form of crosswind aerodynamic loading can be generated in an entirely different manner. Long, slender structures in relatively nonturbulent wind fields can display wake patterns with regular, alternating vortexes. These vortexes, or "eddies" in the flow, have rotational axes parallel to the long axis of the structure. The formation of a vortex from one side of the structure results in a flow field in the neighborhood of the structure that is very asymmetric, even though the shape of the structure may not be asymmetric. As discussed, the asymmetric flow field will cause a net force on the structure in the across-wind direction. This condition is temporary, however. The development of a complementary vortex on the opposite side of the structure will displace the former vortex and cause the net crosswind force to change directions. Crosswind forces may cancel each other out when considering the time-averaged effect, even though the alternating forces are significant. The frequency of the vortex shedding pattern is governed by the dimensions of the cross section, the shape of the cross section, and the wind speed. The Strouhal number is a nondimensional parameter that represents the interaction of these effects:

$$St = \frac{n \cdot D}{U} \tag{2-3}$$

where

St = Strouhal number,
n = Frequency of vortex shedding,
D = Cross-sectional dimension, and
U = Wind speed.

If the vortex shedding frequency is near one of the natural frequencies of the structure, an unfavorable amplification of aerodynamic loading can occur. Steel stacks are sometimes susceptible to this action, and for this reason many are outfitted with helical "strakes" in the upper portion of the structure. Helical strakes disrupt formation and longitudinal coherence of the alternating vortexes, thus mitigating the dynamic crosswind loading. The presence of platforms on vertical vessels and towers may also help prevent the formation of these vortexes. However, problematic crosswind effects have been observed in pairs of vertical vessels (see Section 2.5). These features increase the projected area of the structure and can change the flow pattern such that along-wind forces are increased, but many times this is a worthwhile design concession if resonant oscillations can be avoided.

2.3 AERODYNAMICS OF OPEN-FRAME STRUCTURES

Open-frame structures are common in petrochemical facilities. These structures are characterized by regular bays of open structural framing, the presence of equipment housed within the envelope of the structure, the presence of piping, and other appurtenances such as stairs and handrails. The estimation of wind loads for these structures is complicated by the interactions of the wind with these features.

The wind loads for the frame itself are mainly a function of the solidity of the frame (how much solid area is projected toward the wind), the spacing of the frames, and the total number of frames in the along-wind direction. Upwind frames shield downwind frames to a large degree. Therefore, it is overly conservative to calculate the wind load for one frame and multiply this force by the number of frames. The maximum wind force for an open-frame structure occurs when the wind approaches the structure at a skewed angle to the axis of the structure (Figure 5-2). When the wind is oriented along the structure's axis, leeward frames are more directly shielded by their upwind counterparts. When the wind is skewed to the axis of the structure, leeward columns are more exposed to the wind. These downstream members still do not receive the same wind load as the upwind frame because some of the momentum in the flow has been redirected by the first frame line.

The equipment elements housed within an open-frame structure contribute to the overall wind load on a structure, and the individual supports for these elements must also be designed to resist wind forces. These elements are typically cylindrical shapes (such as vessels or exchangers). The wind force coefficients for these elements are well-known. However, the equipment will experience shielding from the structural frame, thus reducing the wind load from that of an isolated element. In turn, the equipment shields downwind components of the frame, further reducing the forces from what would otherwise be estimated.

Chapter 5 provides recommended methods for estimating the wind forces for open-frame structures and the equipment housed within. There have been some updates and an addition to the methodology that were given in the previous version of *Wind Loads for Petrochemical and Other Industrial Facilities* (ASCE 2011):

- Contribution to wind loading of diagonal bracing that is located in a plane parallel to the nominal wind direction is recognized.
- Wind load–reducing effects of solid floors have been confirmed through research, and the guide contains provisions for taking advantage of this effect.
- A more accurate (and less conservative) method for estimating the shielding of equipment within open-frame structures has been discovered by recent research.
- An alternative analysis technique for estimating the wind loads for highly solid open-frame structures is presented. This method may be particularly useful for estimating envelope wind loads for structures with equipment layouts that have not yet been determined or are expected to change during the life of the structure.

Chapter 6 includes worked examples for open-frame structures demonstrating the application of these updated provisions.

In general, it is accepted to design a pipe rack structure as an open-frame structure with additional loads for pipes and cable trays. It was the experience of the second committee (and this committee concurs) that pipe friction caused by temperature changes and anchor loads govern the longitudinal design loads of a typical pipe rack. Water hammer and other transient loads may also govern the longitudinal forces in pipe racks and bridges. However, the engineer should exercise judgment to ensure longitudinal wind loads are considered for unusual pipe rack configurations, for example, when there is a significant change in elevation of the piping at a road crossing.

There were no design guidelines for pipe bridges in the first edition. the following book (ASCE 2011) added a recommendation to apply wind load on the bridge framing members in a manner similar to the application of wind load on open-frame structures and to apply wind load on piping in a manner similar to that applied on piping in pipe racks.

2.4 AERODYNAMICS OF PARTIALLY CLAD STRUCTURES

Structures with partial cladding are sometimes found in petrochemical or process industry facilities. Because process structures are not usually intended for human occupancy, the placement of cladding may not be complete and may be placed to protect only certain equipment, process areas, or individual substructures (e.g., stairwells). This section discusses some general aerodynamic characteristics of structures with partial cladding. The material in this section is based on research by Amoroso et al. (2010). This research considered rectangular structures with no walls clad, some walls clad (in various configurations), and all walls clad. The scenarios represented in this research did not consider all the cases of partial cladding that may be encountered in a petrochemical facility. For example, compressor shelters, which typically consist of steel portal frame buildings with cladding only on the roof and upper portions of the walls, are not covered in this section.

Clearly, the wind effects for the limiting cases of no cladding and full cladding are covered by considering the structure as an open-frame structure or a fully clad building, respectively. Section 2.3 discussed the general aerodynamics of open-frame structures, and Chapter 5 contains provisions for estimating wind forces for such structures. ASCE 7 includes pressure coefficients for fully clad buildings.

For rectangular buildings with one, two, or three walls clad, the development of forces depends strongly on the wind direction. For the case of one wall clad, the force coefficients are similar to fully clad buildings when the wind direction is nominally perpendicular to the clad side. When the wind direction is nominally parallel to the clad face, the force coefficients are similar to those of an open-frame structure.

Two scenarios are possible for the case of two walls clad: (1) two parallel walls are clad, and (2) two adjacent walls are clad. An important observation from the research was that both configurations cause significantly higher wind forces than the fully clad case. Furthermore, the range of wind directions for which high force coefficients exist extends beyond what is typical of enclosed buildings. For this reason, a designer should consider load cases in which high wind forces act simultaneously along both structure axes.

When two parallel walls are clad, the highest force acts in the direction normal to the clad walls. However, the wind direction is from an angle substantially rotated from the orientation of the force. As the wind approaches the structure, the flow is redirected along a path between the parallel clad walls. The flow is forced to change direction by the presence of the walls, and this change in momentum in the fluid flow is accompanied by a corresponding reaction in the structure. The walls function like vanes in a duct.

When two perpendicular walls are clad, the aerodynamics are quite complex. The maximum forces occur when the clad walls are "cupped" into the wind. However, the maximum force for a given structural axis occurs when the open face is generally oriented to the wind. When the clad faces are not "cupped" into the wind, the development of forces is similar to that on an enclosed building.

For the case of three walls clad, the wind forces are slightly higher than those of an enclosed building when the unclad wall is positioned on the windward face of the structure. Furthermore, high force coefficients persist for a greater range of wind angles than for the fully clad case. When the unclad wall is on the leeward side of the building, the force coefficients are similar to the fully clad case.

Provisions have been included in Chapter 5 for estimating wind forces for partially clad structures. A worked example is also included in Chapter 6 illustrating the application of the provisions. These provisions are for the estimation of overall forces to be resisted by the main wind-force-resisting system. Component and cladding pressures are not provided. Given the high overall force coefficients that can develop on partially clad structures, a designer should exercise caution in using the component and cladding pressures given in ASCE 7 for application on partially clad structures.

However, structures with partial cladding can take a variety of forms in the context of a petrochemical or other industrial facility. Other common structure types include portal-framed metal buildings with cladding only covering the roof and portions of the side and end walls. This structure type is commonly a low-rise building and may be used as a shelter for compressors or other equipment. The 2010 supplement to the 2006 *Metal Building Systems Manual* (MBMA 2010) included provisions for estimating lateral forces for open-sided, multiple-bay, portal-framed buildings. These provisions are based on recent research conducted at the Boundary Layer Wind Tunnel Laboratory at the University of Western Ontario (Kopp et al. 2010).

2.5 AERODYNAMICS OF VERTICAL VESSELS

2.5.1 Cylinder Aerodynamics

Vertical vessels are essentially large circular cylinders, and therefore cylinder aerodynamics form the basis for estimating the wind loads on these types of structures. Force coefficients for circular cylinders in cross flow are well documented in the fluid mechanics literature. ASCE 7 provides force coefficient values for circular cylinders in Figure 29.4-1. As reflected in that figure, the force coefficient for circular cylinders depends on three primary variables: (1) the Reynolds number, (2) the surface roughness of the cylinder, and (3) the aspect ratio, which is the ratio of the cylinder length to the cylinder diameter.

2.5.2 Reynolds Number

The Reynolds number, *Re*, is the ratio of inertial forces to the viscous forces in a flow and was defined in Section 2.2. The aerodynamic force coefficient for circular cylinders is famously sensitive to *Re*. The reason for this sensitivity is the sudden transition to turbulence the small surface boundary layer on a cylinder encounters at $Re \approx 2.5 \times 10^5$ (for smooth cylinders). Turbulent boundary layers have a different velocity profile shape than laminar boundary layers. The presence of turbulence enables momentum to be transferred across the shear layer convectively as well as viscously. This momentum transfer causes higher flow velocities to be present much closer to the cylinder surface in a turbulent boundary layer.

The increased momentum in a turbulent boundary layer enables it to remain attached to the cylinder surface much longer in the presence of the increasingly adverse pressure gradient that develops as the cylinder surface curves away from the flow direction on the downstream side. Prior to the boundary layer's transition to turbulence, the flow separates from a cylinder surface upstream of the centerline, whereas a turbulent surface boundary layer separates downstream of the cylinder centerline. This difference in separation points affects the width of the wake and the width of the negative pressure zone on the downstream side of the cylinder. This zone is narrower when the flow separation is delayed until after the boundary layer has passed the cylinder centerline, and consequently the aerodynamic drag is lower. Vertical vessels are large enough that the *Re* for realistic atmospheric flow conditions will be approximately equal to 10^7, which exceeds this critical regime. Smaller elements, such as small-diameter pipes and cylindrical rail elements, must be considered more closely.

2.5.3 Surface Roughness

Aerodynamic drag can be generated by two general mechanisms: form drag and friction drag. Form drag is caused by the development of pressures that act normal to the surface of a body, and this type of drag is typically dominant for bluff bodies such as cylinders. Friction drag is caused by the shear stresses parallel to the surface of a body that develop as a fluid passes over the body's surface. Friction drag is more important for "streamlined" shapes, such as airplane wings, which do not produce large regions of separated flow (or wakes).

The presence of surface roughness on a cylinder can cause the surface boundary layer to become turbulent at lower values of *Re* than would be expected for a smooth cylinder surface. Consequently, the form drag would be lower for a rougher cylinder than for a smooth cylinder. However, there is a competing mechanism that mitigates this effect. The increased surface roughness increases the friction drag. So, even though lower drag coefficients can exist at lower values of *Re*, the ultimate drag coefficient at high *Re* is higher for rougher surfaces.

2.5.4 Aspect Ratio

The aspect ratio for a vertical vessel is defined as the height divided by the diameter. Reducing the aspect ratio for a circular cylinder causes a reduction in the aerodynamic force coefficient. The mean flow field for an infinitely long cylinder is two-dimensional. Finite length introduces three-dimensionality into the flow field. In the case of a finite-length cylinder, the flow has an alternative path around the cylinder—around the ends rather than solely around the circumference.

This additional flow path reduces the quantity of flow that travels around the circumference on a per-length basis. The reduced quantity of flow translates into lower pressure on the circumferential surface of the cylinder.

For vertical vessels at grade, only the top end is effective in providing an alternative flow path. This orientation is similar to a chimney or a stack. ASCE 7 provides force coefficients for these types of structures that consider the effects of finite length, and Chapter 5 of this book recommends the use of these values for vertical vessels. If a vessel is not oriented with one end at grade, then the use of force coefficients for ground-mounted cylindrical structures in ASCE 7 will likely result in conservative estimates of the wind load.

2.5.5 Treatment of Appurtenances

Vertical vessels in real industrial settings are not ideal, smooth cylinders. Most vertical vessels have appurtenances such as ladders, handrails, platforms, and piping of various diameters on their surfaces. The presence of these elements complicates the flow field and the calculation of wind loading. When the details of these elements are unknown (e.g., during preliminary analysis or design), the provisions of Chapter 5 suggest accounting for their contribution to the wind load by augmenting the dimensions of the vessel itself. When more details are known, the contribution of these elements to the total wind load can be calculated by considering the projected areas and corresponding force coefficients of the elements. Considerable shielding will occur, and it is recommended in the provisions of Chapter 5 to only consider the net projected area of elements on the windward side of the vessel. An exception is for handrails on platforms, which can have sufficient separation to preclude shielding.

2.5.6 Neighboring Vessels

It is common that vertical vessels are arranged closely as pairs or in other groupings. An example of such a pairing is a fluid catalytic cracking reactor and a catalyst regenerator. When vessels are closely arranged, the aerodynamics can be considerably more complicated. For some wind directions, an upstream vessel will shield the neighboring vessel, thus reducing the wind load on the downstream vessel. When vessels are oriented side by side with respect to approaching wind, their proximity can cause a local acceleration of the wind between the two vessels, thus increasing the wind loads on both structures. Chapter 5 suggests increasing the force coefficient for these crosswind arrangements. However, there is no provision in Chapter 5 to account for the possibility of shielding, which is consistent with the provisions of ASCE 7.

2.5.7 Directionality

The radial symmetry of single vertical vessels increases the likelihood that the worst wind direction will coincide with the critical aerodynamic orientation. However, for pairs or groups of vertical vessels, this radial symmetry is disrupted.

In these scenarios, the designer may choose to use a lower value of the directionality factor, K_d.

2.5.8 Gust Effects

Vertical vessels can be quite slender, and they can be quite massive when filled with fluid. As such, they may have low fundamental vibration frequencies. The reduced natural frequency increases the likelihood that along-wind buffeting from the wind will amplify the wind loads. This effect is represented in the calculation of the gust effect factor, G. The example calculation in Chapter 6 illustrating the estimation of wind loads for vertical vessels demonstrates the approximate magnitude this type of dynamic load amplification can produce. In this example, the calculated values of G for the rigid and flexible cases are 0.85 and 1.198, respectively, corresponding to a dynamic load amplification of approximately 41% when the vessel is filled with fluid.

2.5.9 Vortex-Induced Vibration

Members of the task committee have observed problematic occurrences of vortex-induced vibrations for pairs of tall, slender vertical vessels situated near one another. The phenomenon occurred during relatively low wind speeds and when the direction of the wind aligned with the centers of the two vertical vessels. That is, there was one windward vessel and one leeward vessel. Large, resonant crosswind vibrations developed in both vessels, with observed movements being higher in the leeward vessel.

2.6 OTHER WIND-LOADING CODES, STANDARDS, AND GUIDES

In the United States, ASCE 7 governs the estimation of wind loads for analysis and design purposes (except for bridges, which are covered by AASHTO LRFD). ASCE 7 is incorporated by reference into the International Building Code. This publication, *Wind Loads for Petrochemical and Other Industrial Facilities*, is intended to supplement, and be compatible with, the wind load provisions of ASCE 7. However, the committee responsible for this book recognizes that it does not cover all the possible structure types or forms that may be encountered in the process industries. Members of this committee have found the following documents and resources useful because they contain a wide variety of wind loading shape coefficients:

- AS/NZS 1170.2, *Structural Design Actions—Part 2: Wind Actions* (Standards Australia 2011).
- *Eurocode 1: Actions on Structure—Part 1-4: General Actions—Wind Actions* (CEN 2010).

- ESDU Wind Engineering Series. ESDU (formerly the Engineering Sciences Data Unit) is a subscription-based service that provides access to engineering data and software.

Wind load provisions contained in these resources may have a different basis than that of ASCE 7 and this book. The analyst or designer should be careful to make sure the use of various formulas and coefficients is compatible with other aspects of the analysis methodology (e.g., wind speed averaging times, mean versus peak effects).

2.7 RESEARCH PROGRESS AND FUTURE NEEDS

The first version of the report (ASCE 1997) reviewed the analysis practices of several companies and provided consensus guide provisions for the estimation of wind loads. For the most part, the resulting provisions were based on extensions of methods from ASCE 7 and engineering judgment rather than on research. Recognizing this limitation with the guide provisions, the committee identified several research priorities. Since then, progress has been made in addressing some of these issues.

- Force coefficients and shielding effects for pipes in a pipe rack structure were studied in a wind tunnel by Liu et al. (2008). The effects of shielding were found to be dependent on the number of cylinders in each pipe group, their spacing, and the turbulence in the flow. The wind tunnel experiments were in relatively low Reynolds numbers, so further research is required to more fully define the aerodynamics of pipe racks at full scale.
- Insight into the shielding of equipment inside open-frame structures has been sought out in the literature and studied in a wind tunnel (Amoroso and Levitan 2009). It has been found that the force reductions for equipment inside these structures are generally greater than the provisions the previous edition of this publication suggested.
- The question of an upper-bound force coefficient for open-frame structures has also been investigated, and it has been found that no such limit exists for unlimited plan geometries. However, if the envelope dimensions of an open-frame structure are known, a reasonable upper-bound force coefficient can be estimated (Amoroso and Levitan 2011).
- Limited experimental data are now available concerning the influence of a large-diameter vertical pipe adjacent to a vertical vessel. It has been observed that aerodynamic interference between these two elements can sometimes amplify the wind loads (Amoroso 2007).
- The effectiveness of simplified methods for estimating the contributions of ladders, nozzles, and small pipes to vertical vessel wind loads by using an increased vessel diameter has been evaluated for a particular case

(Amoroso 2007, Amoroso and Levitan 2009). It was found that the simplified method recommended by the previous edition of this publication was sometimes not conservative.

- The presence of solid flooring has been found to reduce the mean wind loads for open-frame structures (Amoroso and Levitan 2009).
- Petersen (1997) studied the wind environment for models of refineries in wind tunnel experiments. The motivation for the research was to improve the analysis of the mixing and dispersion of pollutants in the atmosphere. The experiments showed that the surface roughness inside the plants corresponded to suburban and dense urban terrain.

Despite this progress, many of the research questions originally identified in the first edition (ASCE 1997) have not been addressed to the knowledge of this committee. These are

- What are appropriate force coefficients and shielding effects for cable trays in a pipe rack structure?
- What are the loads on vertical rows of pipes (force coefficients for wind directions from perpendicular to parallel to the row of pipes)?
- What force coefficients should be used for different sizes, types, and orientations of platforms on a vessel?
- Although a limited case study has been performed, a general understanding of the appropriate force coefficients for ladders, nozzles, and small pipes remains. Can these items be accounted for as proposed in the Chapter 5 recommendations?
- What are the effects of irregular (nonrectangular) plan view open-frame structures on the magnitude and application of the wind-induced force?

In addition to these outstanding research questions, discussions among committee members have identified the following items for consideration by researchers:

- How does the currently available research regarding pipe shielding compare with results from high Reynolds number experiments?
- What are the longitudinal wind forces on pipes in a pipe rack structure?
- Is the gust effect factor for rigid buildings as specified by ASCE 7 applicable for rigid open-frame structures?
- Is the gust effect factor formula for flexible structures specified by ASCE 7 applicable for tall, flexible industrial structures, such as trussed towers, vertical vessels, or stacks?
- Although more refinement has been achieved in estimating the shielding effects for equipment inside open-frame structures, the shielding estimates are gross quantities applied uniformly to all equipment items on a given story. What are the wind loads for individual pieces of equipment?

- What are the components and cladding pressures for partially clad structures?
- How do the recommendations for estimating wind loads for air coolers compare to wind tunnel test results?

References

AASHTO. 2017. *LRFD bridge design specifications*. Washington, DC: AASHTO.

Amoroso, S. 2007. "Wind loads for petrochemical structures." Ph.D. dissertation, Dept. of Civil and Environmental Engineering, Louisiana State Univ.

Amoroso, S., K. Hebert, and M. Levitan. 2010. "Wind tunnel tests for mean wind loads on partially clad structures." *J. Wind Eng. Ind. Aerodyn.* **98** (12): 689–700. https://doi.org/10.1016/j.jweia.2009.08.009.

Amoroso, S., and M. Levitan. 2009. "Recent research for wind loads on petrochemical structures." In *Proc., ASCE/SEI 2009 Structures Congress*, Austin, TX, 2088–2097.

Amoroso, S., and M. Levitan. 2011. "Wind loads for high-solidity open-frame structures." *Wind Struct.* **14** (1): 1–14. https://doi.org/10.12989/was.2011.14.1.001.

ASCE. 1997. *Wind loads and anchor bolt design for petrochemical facilities*. Reston, VA: ASCE.

ASCE. 2011. *Wind loads for petrochemical and other industrial facilities*. Reston, VA: ASCE.

ASCE. 2017. *Minimum design loads and associated criteria for buildings and other structures*. ASCE 7-16. Reston, VA: ASCE.

CEN (European Committee for Standardization). 2010. *Eurocode 1: Actions on structures—Part 1-4: General actions—Wind actions*. Brussels, Belgium: CEN.

ICC (International Code Council). 2018. *International building code (IBC)*. Washington, DC: ICC.

Kopp, G. A., J. K. Galsworthy, and J. H. Oh. 2010. "Horizontal wind loads on open-frame, low-rise buildings." *J. Struct. Eng.* **136** (1): 98–105. https://doi.org/10.1061/(ASCE)ST.1943-541X.0000082.

Liu, X., M. Levitan, and D. Nikitopoulos. 2008. "Wind tunnel tests for mean drag and lift coefficients on multiple circular cylinders arranged in-line." *J. Wind Eng. Ind. Aerodyn.* **96** (6–7): 831–839. https://doi.org/10.1016/j.jweia.2007.06.011.

MBMA (Metal Building Manufacturer's Association). 2010. *Metal buildings systems manual: 2010 supplement to the 2006 edition*. Cleveland, OH: MBMA.

Petersen, R. L. 1997. "A wind tunnel evaluation of methods for estimating surface roughness length at industrial facilities." *Atmos. Environ.* **31** (1): 45–57. https://doi.org/10.1016/S1352-2310(96)00154-9.

Standards Australia. 2011. *Structural design actions—Part 2: Wind actions*. AS/NZS 1170.2:2011. Sydney, Australia: Standards Australia.

CHAPTER 3

Review of Existing Design Practices

3.1 INTRODUCTION

The ASCE Petrochemical Committee responsible for the first publication of this document in 1997 conducted a review of design practices within the engineering community at that time (mid-1990s) to address the variable nature of such practices for industrial structures. Thirteen design practices for estimating wind loads on pipe racks, open structures, and pressure vessels were reviewed. The results of that effort were presented in the first edition.

Section 3.2 presents the results of a survey conducted during the preparation of the second book to complement the efforts of the first committee.

Section 3.3 provides a discussion of design practices for structural types that are already covered by other industry standards.

3.2 SURVEY OF EXISTING PRACTICES AND IMPACTS OF FIRST EDITION OF GUIDELINES

The 1997 ASCE task committee report, *Wind Loads and Anchor Bolt Design for Petrochemical Facilities*, attempted to address the variable nature of design practices for industrial structures within the engineering community (ASCE 1997). The second committee conducted a web-based survey of design engineers and engineering managers from large engineering consulting firms to assess the level of acceptance and effectiveness of the 1997 publication and to determine future document improvement needs.

The level of acceptance and effectiveness of the original 1997 book can be represented by the following survey results: (1) 80% of the respondents were knowledgeable of the document, and (2) 57% of the respondents referenced the publication in their current company design specifications.

In referencing the 1997 book, company specifications have

- Modified the tributary areas of pipes on racks,
- Reduced wind loading on piping in a pipe rack owing to multiple pipe rack levels, and
- Used an alternative force coefficient for cable trays.

The committee wishes to thank those who responded to the survey. The information gathered has and will continue to guide current and future committees in their efforts to provide for the needs of the petrochemical and industrial engineering community. Table 3-1 provides the responses to each question and the desired enhancements for the document. The committee attempted to address the desired enhancements conveyed through the survey results, and some are reflected in these chapters.

3.3 EXISTING DESIGN PRACTICES FOR STRUCTURES NOT COVERED BY THIS GUIDE

This section addresses the design practices for calculating the wind loads for tanks, steel stacks, and cooling towers.

3.3.1 Tanks

3.3.1.1 Overview

Most storage tanks located in petrochemical facilities are fabricated from steel and conform to one of the following documents:

- API 650, *Welded Steel Tanks for Oil Storage* (API 2013b);
- API 620, *Design and Construction of Large, Welded, Low-Pressure Storage Tanks* (API 2013a);
- AWWA D100, *Welded Carbon Steel Tanks for Water Storage* (AWWA 2011);
- UL-142, *Standard for Steel Aboveground Tanks for Flammable and Combustible Liquids* (UL 2019);
- NFPA 22, *Standard for Water Tanks for Private Fire Protection* (NFPA 2018c);
- NFPA 30, *Flammable and Combustible Liquids Code* (NFPA 2018b), and
- NFPA 30A, *Code for Motor Fuel Dispensing Facilities and Repair Garages* (NFPA 2018a).

With respect to wind loads, the documents offer the following guidelines.

Table 3-1. Survey Responses.

Question	*Responses*	
How familiar are you with the 1997 ASCE task committee report *Wind Loads and Anchor Bolt Design for Petrochemical Facilities* for application of wind loads?	Not familiar	21%
	Somewhat familiar	25%
	Very familiar	54%
Do your company specifications reference the 1997 ASCE task committee report *Wind Loads and Anchor Bolt Design for Petrochemical Facilities* for application of wind loads?	Unknown	14%
	No	29%
	Yes	57%
Does your firm reduce the wind loading on the piping in a pipe rack owing to multiple pipe rack levels?	Unknown	23%
	No	65%
	Yes	12%
Does your firm consider longitudinal loading on pipe racks?	Unknown	19%
	No	15%
	Yes	67%
Does your firm use an upper limit for evaluation of open-frame structures (e.g., consider the load on a fully clad structure with the same envelope shape)?	Unknown	26%
	No	56%
	Yes, specify	19%
	Specified responses included the following comments: • Varies among projects—upper limit as fully clad. • Enclosed structure (C_f = 1.3). • Yes, not to exceed load of fully clad structure. • Fully clad structure with C_f = 1.3.	
Would you rather have recommendations that require more detailed/complicated (and perhaps more accurate) analysis methods or easier-to-use estimates that may provide more conservatism?	More detailed	8%
	Easier to use	81%
	Other, please elaborate	12%
	Elaborated responses included the following comments: • Both, so the engineer can choose. • Keep it simple but provide examples and guidance. • More accurate analysis methods.	

(Continued)

Table 3-1. Survey Responses. (Continued)

Question	*Responses*		
Are there specific recommendations or modifications that you feel should be made to the *Wind Loads on Petrochemical Facilities* publication or in the presentation of the recommendations? If so, please describe the modifications you feel would make the recommendations better and more usable to the engineering community.	Unknown No Yes		19% 15% 67%
How beneficial would new recommendations from the ASCE Petrochemical Committee be for the following structures?	Very	Somewhat	Not
Arrays of structures and adjacent structures	62%	25%	12%
Compressor sheds	41%	41%	18%
Structures with dense equipment	64%	32%	4%
One-bay structures	52%	26%	22%
Stacks	58%	38%	4%
Partially clad structures	55%	41%	5%
Cooling towers	58%	33%	8%
Fin-fan-type coolers	36%	50%	14%
Pipe bridges	76%	20%	4%
Are there any other structure types that you feel should be evaluated by the ASCE Petrochemical Committee? Which specific structure types?	Specified responses included the following comments: • Miscellaneous pipe supports. • Equipment foundations (design and transfer of wind loads to foundations at grade). • Offshore structures. • More guidance on when shielding can be employed and limitations of shielding for wind loading. • Office/administration buildings and control room structures. • Sloping conveyor structures, especially in the longitudinal direction. • Elevated enclosures where the floor and roof are exposed to wind.		

Table 3-1. Survey Responses. (Continued)

Question	*Responses*

i. 81% indicated a preference for easier-to-use methods that may provide more conservatism versus more detailed analysis or providing both options in the document.
ii. 67% consider longitudinal loading on pipe racks.
iii. More examples or structural types (see Note vi).
iv. Guidance for shielding.
v. Inclusion of design guidance for offshore structures.
vi. List includes pipe bridges, structures with dense equipment, arrays of all industrial and not specifically petrochemical facilities.

3.3.1.2 API 650

Section 5.2.1k of API 650 contains provisions for the selection of design wind speeds and pressures for oil storage tanks based on risk category. As an alternative, pressures may be determined in accordance with ASCE 7 (exposure category and risk category provided by purchaser) or the national standard for the specific conditions for the tank being designed.

3.3.1.3 Uplift Forces per API 650

Section 5.2.1k of API 650 limits the design uplift pressure on the roof (wind plus internal pressure) relative to 1.6 times the design pressure determined in Section F.4.1. It goes on to stipulate that windward and leeward horizontal wind loads on the roof are conservatively equal and opposite and therefore are not included in these pressures.

A survey of the literature indicates that roof sheet damage to tanks is common; however, when this occurs, the failure is similar to an internal overpressure failure of a tank, where the roof-to-roof angle fails. This in turns allows the stiffening angle or wind girder to stay in place, and thus the contents of the tank are usually not released. In the case of external floating roofs, experience has shown that because of unsymmetrical loadings on the roof, the floating roof may tend to shift. In a combined rain and wind event, there have been instances where this shift resulted in sufficient accumulation of rainwater away from the drain, resulting in the roof sinking (roof failure).

3.3.1.4 Wind Load on Tanks (Overturning Stability)

API 650 Section 5.11 provides analytical methods to determine stability for tankage. This is primarily a concern for small tanks. Section 5.12 of API 650 provides prescriptive methods to calculate the requirements for anchorage.

3.3.1.5 API 620

Section 5.4.1k of API 620 specifies the selection of design wind speeds and surface pressures. The design wind speed (V) shall be the 3 s gust design wind speed (mi/h) determined from ASCE 7, Figures 26.5-1 and 26.5-2 or the 3 s gust design wind speed specified by the purchaser. For tank components exposed to wind up to 80 ft aboveground, API 620 tabulates average and maximum pressures, normalized for design wind speed. For tank components located more than 80 ft aboveground, use ASCE 7-16 to determine wind pressures.

Alternatively, pressures may be determined in accordance with ASCE 7 or a national standard for the specific conditions for the tank being designed.

Average wind pressure on the roof shall be used to design the roof-to-shell compression region and for overturning. Maximum wind pressure shall be used to design the roof and shell.

3.3.1.6 AWWA D100

AWWA D100 offers the following regarding wind loads:

Wind pressure shall be calculated by the following formula from AWWA D100:

$$P_w = q_z G C_f \geq 30\, C_f \tag{3-1 AWWA}$$

where

P_w = Wind pressure applied to horizontal projected area (psf);

G = Gust effect factor, which shall be taken as 1.0 or may be calculated using the procedure given in ASCE 7; the calculated gust effect factor shall be based on a damping ratio of 0.05 and shall not be less than 0.85;

C_f = Force coefficient; and

q_z = Velocity pressure evaluated at height z of the centroid of the projected area (psf)

$$q_z = 0.00256\, K_z I\, V^2 \tag{3-2 AWWA}$$

where K_z is the velocity pressure exposure coefficient evaluated at height z of the centroid of the projected area, and z is the height above the finished grade.

Basic wind speeds are based on a 3 s guest speed at 33 ft (10 m) aboveground and an annual probability of 0.02 of being equal or exceeded (50-year mean recurrence interval). In special wind regions, tanks may be exposed to greater wind speeds.. In such cases, the basic wind speed shall be specified by the Authority Having Jurisdiction.

Velocity pressure coefficients are provided for Exposure C and Exposure D in AWWA D100. Exposure C shall be used unless otherwise specified. The velocity pressure coefficient shall be evaluated at height z of the centroid of the project wind area. For intermediate heights, use linear interpolation of the velocity pressure coefficients.

Chapter 3 of AWWA D100 was updated in 2011 to match the wind exposure definitions in ASCE 7-05 (2005). The significant changes to the definitions of basic wind speeds represented in the change from ASCE 7-05 to ASCE 7-10 (2010) are not represented in AWWA D100-11.

3.3.2 Steel Stacks

Most stacks located in petrochemical facilities are steel stacks. There are two basic types of steel stacks: (1) self-supporting stacks together with the foundation would remain stable under all working conditions without additional support, and (2) guyed or braced stacks; not all external applied loads are carried by the stack shell and therefore guys or braces are provided to ensure stability.

ASME STS-1-2016 (2016) and *Model Code for Steel Chimneys* (CICIND 2010) provide the requirements and guidelines for design, fabrication, erection, and maintenance of steel stacks and their appurtenances. With respect to wind loads, steel stacks should be designed to resist the wind forces in both along-wind and crosswind directions. In addition, the variation of pressure along the circumference of the shell should be considered. Section 4.3.3.2 of ASME STS-1-2016 provides methods for accommodating both ASCE 7-05 and 7-10 type wind speeds.

3.3.2.1 Along-Wind Loads

ASME STS-1-2016 requires that wind loads on steel stacks along the wind direction be calculated as the sum of two components, one caused by a mean wind speed and the other caused by fluctuating wind gusts. The dynamic response is accounted for using a modified gust factor (similar to the gust factor for flexible or dynamically sensitive structures as described in ASCE 7, Section 26.11.5) depending on the natural frequency and the geometric properties of the steel stack. A similar along-wind load provision is included in *Model Code for Steel Chimneys*.

Per Section 4.3.3.1 of ASME STS-1-2016, the design wind loads in the along-wind direction are given by

$$w(z) = \overline{w}(z) + w_D(z)$$

and

$$\overline{w}(z) = \frac{C_f q_z D}{12(1 + 6.8 I_{\overline{Z}})}$$

$$w_D(z) = \frac{3zM_0}{h^3}\left[G_f(1 + 6.8 I_{\overline{Z}}) - 1\right]$$

where

$w(z)$ = Total along-wind load on stack per unit height (lbf/ft),
$\overline{w}(z)$ = Mean along-wind load on stack per unit height (lbf/ft),

$w_D(z)$ = Fluctuating along-wind load on stack per unit height (lbf/ft),
C_f = Force coefficient given in the Mandatory Appendix of ASME STS-1-2016 (or Figure 29.4-1 of ASCE 7-16),
q_z = Velocity pressure, calculated using Equation (4-4) of ASME STS-1-2016 (lbf/ft),
D = Diameter of stack at elevation under consideration (in.),
$I_{\overline{Z}}$ = Intensity of turbulence at height z, calculated according to the Mandatory Appendix of ASME STS-1-2016 [or Equation (26.11-7) of ASCE 7],
z = Elevation under consideration (ft),
h = Height of stack (ft),
M_0 = Moment at the base of the stack owing to $\overline{w}(z)$ loading (lbf-ft), and
G_f = Gust effect factor, calculated according to the Mandatory Appendix of ASME STS-1-2016 (or Section 26.11.5 of ASCE 7).

3.3.2.2 Crosswind Loads from Vortex Shedding Loads

When stacks are subjected to a steady wind, the periodic shedding of vortices will cause swaying oscillations in a direction normal to that of the wind. If the vortex shedding frequency is resonant with the natural frequency of the stack, it may result in large vibrations. Section 5.2.2 (a) of ASME STS-1-2016 can be used to evaluate if vortex shedding needs to be considered. Example calculations in Nonmandatory Appendix E of ASME STS-1-2016 may be followed if needed. Per Section 5.2.2 (a) of ASME STS-1-2016, fatigue analysis must be considered because of vortex shedding loads under certain conditions.

3.3.2.3 Ovalling

In addition to transverse swaying oscillations, a steel stack may also be subjected to ovalling. Ovalling results from positive pressure on the upwind side of the stack and negative pressure on the sides and back. Ovalling resonance in the circular cross-sectional plane occurs as a result of vortex excitation. A lined stack is more resistant to ovalling because the lining contributes to the high natural frequency and increased damping for the elastic ring. Therefore, ovalling need not be considered for lined stacks. Unlined stacks possess very little damping to restrict ovalling and may experience excessive stresses and deflections at the critical ovalling wind velocity. Per ASME STS-1-2016, it may be assumed that the unlined stack is prone to ovalling vibrations if the critical wind velocity for ovalling is less than or equal to the mean hourly wind speed. A simple practical cure against ovalling is to use stiffening rings.

3.3.3 Cooling Towers

Most cooling towers located in petrochemical facilities are wood-frame structures constructed on site with corrugated cladding, wooden or fiber-reinforced polymer (FRP) louvers, and FRP fan cylinders (or shrouds). They typically are at or near grade, less than 60 ft tall, rectangular in shape, and have a modest inward slope on two opposite sides. In some cases, cooling tower structural systems are constructed of materials other than wood (FRP or reinforced concrete) to increase their durability or to provide portability (metal). The Cooling Technology Institute (CTI) Standard STD-114 (2014) provides design guidance for wooden cooling towers.

Prior to the creation of this standard, unless specified otherwise by the owner, the manufacturer established the wind load criteria for a given tower design. In such instances, it is not always clear how the design wind pressures were determined. For example, a manufacturer may have simply stated that the basic design criteria shall be a 30 psf wind load. Closer examination of the application of such design values reveals that, although they may be suitable for calculating loads on the main structural frame [i.e., the main wind-force-resisting system (MWFRS)], these values often underestimate high local wind pressures acting along the corners and edges of the tower frame. Nonetheless, given the current CTI standard, it is now recognized that ASCE 7 is to serve as the design basis for determining wind loads on cooling towers. If applied properly, the CTI standard together with ASCE 7 should enable the designer to account for global and local wind pressures acting on a cooling tower structure.

Kelly (2006) addresses special requirements for cooling towers per the International Building Code (IBC) (2003), ASCE 7-02 (2002), and the Florida Building Code (FBC). Kelly correctly states that the approach taken to calculate wind loads in the IBC and FBC is based on ASCE 7. Kelly also provides examples for each case, demonstrating how to calculate wind loads on a cooling tower, citing the methodology and terminology used in ASCE 7-02. IBC 2003 adopted a methodology similar to Method 1—Simplified Procedure of ASCE 7-02. Kelly lists some, but not all, of the requirements of Section 6.4 of ASCE 7-02, which establishes the basis for using Method 1. As noted by Kelly, Method 2—Analytical Procedure of ASCE 7-02 applies when

- IBC 2003 does not apply,
- The building has a mean roof height of 60 ft or more (i.e., the top of the cooling tower is at a height of 60 ft or more), and
- The building (i.e., cooling tower) does not have a site location for which channeling or buffeting of upwind obstructions warrants special consideration.

The simplified procedures (Method 1) for determining MWFRS and component and cladding (C&C) wind pressures in ASCE 7-02 are comparable to the MWFRS envelope procedure for enclosed simple diaphragm buildings and the C&C simplified provisions for low-rise buildings in Chapters 28 and 30 of ASCE 7-16, respectively.

Sections 28.5.2 and 30.1 of ASCE 7-16 list additional caveats for applying these methods, among them that the building (i.e., cooling tower)

- Is a simple diaphragm building,
- Has a flat roof, gable roof, or hip roof,

- Is enclosed and conforms to the wind-borne debris provisions of Section 26.12.3,
- Is regular shaped,
- Is not classified as a flexible structure,
- Does not have response characteristics making it subject to across-wind loading, vortex shedding, or instability because of galloping or flutter,
- Is approximately symmetrical in cross section in each direction with either a flat roof or a gable or hip roof with 45°, and
- Is exempted from torsional load cases.

The results as summarized by Kelly indicate that ASCE 7-02 provided suitable guidance for the design of cooling towers. Given the structural features of most cooling towers and to reduce the time required to calculate wind pressures, it is recommended that a cooling tower for a petrochemical facility be designed for wind loads using the MWFRS envelope procedure for enclosed simple diaphragm buildings and the C&C simplified provisions for low-rise buildings in Chapters 28 and 30 of ASCE 7-16, respectively. However, if a cooling tower is located on top of a building or other supporting structure in a manner that results in the top of the tower being at a height greater than 60 ft, it is recommended that it be designed for wind loads using other applicable provisions of ASCE 7.

The calculation of wind pressures on fan shrouds at the top of the cooling tower should be treated separately from the calculation of wind pressures for the main cooling tower structure surfaces. The external pressures for these elements may be calculated according to the methods in Section 29.4.2 of ASCE 7-16 for circular bins, silos, and tanks. These provisions include consideration of interference effects associated with groups of tanks with small center-to-center spacings. Because the fan shrouds are open topped, roof pressures need not be considered.

References

API (American Petroleum Institute). 2013a. *Design and construction of large, welded, low-pressure storage tanks, twelfth edition.* API 620. Washington, DC: API.

API. 2013b. *Welded steel tanks for oil storage.* API 650. Washington, DC: API.

ASCE. 1997. *Wind loads and anchor bolt design for petrochemical facilities.* Reston, VA: ASCE.

ASCE. 2005. *Minimum design loads for buildings and other structures.* ASCE 7-05. Reston, VA: ASCE.

ASCE. 2010. *Minimum design loads for buildings and other structures.* ASCE 7-10. Reston, VA: ASCE.

ASCE. 2017. *Minimum design loads and associated criteria for buildings and other structures.* ASCE 7-16. Reston, VA: ASCE.

ASME (American Society of Mechanical Engineers). 2006. *Steel stacks.* ASME STS-1-2006. New York: ASME.

AWWA (American Water Works Association). 2011. *Welded carbon steel tanks for water storage.* ANSI/AWWA D100-11. Denver: AWWA.

CICIND (International Committee on Industrial Chimneys). 2010. *Model code for steel chimneys, revision 1.* Zurich, Switzerland: CICIND.

CTI (Cooling Technology Institute). 2014. *Design of cooling towers with Douglas fir lumber.* STD-114. Houston: CTI.

Kelly, D. S. 2006. "Wind load rated packaged cooling towers." In *Proc., CTI Annual Conf.*, San Diego, CA.

NFPA (National Fire Protection Association). 2018a. *Code for motor fuel dispensing facilities and repair garages.* NFPA 30A. Quincy, MA: NFPA.

NFPA. 2018b. *Flammable and combustible liquids code.* NFPA 30. Quincy, MA: NFPA.

NFPA. 2018c. *Standard for water tanks for private fire protection.* NFPA 22. Quincy, MA: NFPA.

UL (Underwriters Laboratories). 2019. *Standard for steel aboveground tanks for flammable and combustible liquids.* UL 142. Washington, DC: UL.

CHAPTER 4

Recommended Guidelines Part I: Design Considerations and Methods

4.1 INTRODUCTION

Chapter 4 provides guidance for wind load analysis and design of petrochemical and industrial structures and includes a broad range of topics related to wind effects for these types of facilities.

This chapter begins with observations on the historical performance of petrochemical facilities during hurricanes and provides some related general design considerations. Sections 4.3 and 4.4 provide background information on alternate methods for the determination of wind loads, specifically wind tunnel testing and computational fluid dynamics (CFD) approaches. Information is provided in Section 4.5 regarding recommended wind load combinations for petrochemical structures. Special wind load provisions related to US Department of Transportation regulations for liquefied natural gas (LNG) facilities are presented in Section 4.6. Section 4.7 provides references and summaries of current industry guidelines for the evaluation of wind loads on existing structures and foundations. Finally, a discussion of the uncertainty associated with the determination of wind loads is provided in Section 4.8.

4.2 HISTORICAL PERFORMANCE

This section highlights the performance of industrial structures during past hurricane events. Because comprehensive performance data of petrochemical and industrial facilities are not publicly available, it is worth noting that during the last 50 years, many industrial facilities along the Gulf and Atlantic coastlines have experienced at least tropical storms or minor hurricanes and have performed well. An example of good performance can be seen in Figure 4-1, which shows an industrial facility in Gregory, Texas, under construction at the time of Hurricane

Figure 4-1. Active facility construction site located in Gregory, Texas, under construction during Hurricane Harvey that sustained no major damage resulting from hurricane-force winds.

Harvey. However, many facilities have not experienced a design-level event (typically a moderate Category 2 or stronger hurricane), leading some in the industry to believe their facilities can withstand major hurricane-force winds without sustaining significant damage.

Although catastrophic damage to a facility is not normal, Figure 4-2 shows a damaged catalytic unit at a refinery in Corpus Christi, Texas, following Hurricane Celia in 1970. Based on a (blurry) sketch of the unit configuration prior to the event, the unit collapsed at a distance about one-third of its height from its base, with a surge separator, hot bins, and a steam drum toppling to the ground. The unit was noted as being one of the older structures on the site.

Following are observations of damage sustained during past hurricane events. In each case, a storm category is provided at landfall per the National Hurricane Center; however, the wind intensity at the site is unknown.

4.2.1 Storage Tanks

Godoy (2007) indicates that the performance of oil storage tanks was inconsistent. The paper points out that floating roof tanks with wind girders in place fared well, while similar adjacent tanks without wind girders suffered significant damage.

Figure 4-3 shows some of the destruction caused by Hurricane Hugo. Hurricane Hugo passed through the Caribbean, causing billions of dollars of

Figure 4-2. Corpus Christi refinery unit, Hurricane Celia.

Figure 4-3. Petroleum storage tanks, Hurricane Hugo.
Source: Courtesy of NOAA.

damage. Amerada Hess incurred significant expenses repairing and upgrading its St. Croix refinery in the Virgin Islands after it sustained heavy damage from the storm. The National Research Council and J&H Marsh & McLennan (1998)

Figure 4-4. Storage tank under construction, Hurricane Katrina.
Source: Courtesy of LCA.

reported that a large number of petroleum tanks, including 14 of the 500,000- to 600,000 barrel capacity storage tanks, sustained extensive damage at the refinery.

During Hurricanes Katrina and Rita (Category 3), several steel oil storage tanks were severely damaged by wind forces, while numerous other tanks were damaged by flooding after floating away from their original positions (Figures 4-4 through 4-7).

Hurricane Humberto, which occurred in September 2007, was probably the fastest developing storm in recent history. The storm rapidly intensified from a tropical depression to a hurricane within 19 h. The storm inflicted more damage than its Category 1 strength would suggest. It is reported the storm caused damage to an LNG storage vessel in the early construction phase on the Texas–Louisiana coastline. It is suspected the tank was vulnerable because of the construction phase at the time of the event. In addition to the industrial equipment damage, there were numerous power outages that affected several industrial complexes.

4.2.2 Electrical Power Transmission

The poor performance of electrical power transmission and distribution systems can be a major problem at an industrial facility. Calvert and Fouad (2000) point

Figure 4-5. Oil storage tanks, Hurricane Katrina.
Source: Courtesy of NOAA.

Figure 4-6. Tank damage, Hurricane Katrina.
Source: Courtesy of NOAA.

Figure 4-7. Tank insulation and cladding damage, Hurricane Rita.
Source: Courtesy of LCA.

out that the National Electrical Safety Code (IEEE 1996) wind loads applied to power poles are significantly lower than those calculated by ASCE 7.

Hurricane Frances (Category 2) made landfall along the east coast of Florida. It was one of four storms that impacted Florida in 2004. Some buildings were in a temporary state of repair when Frances hit. Prolonged power outages were also experienced during this time (Figures 4-8 through 4-10).

During Hurricanes Katrina and Rita (Category 3), power supplies were severely affected (Figures 4-11 and 4-12); approximately 1 million wooden poles and 300 towers or steel poles collapsed because of wind forces.

4.2.3 Cooling Tower

During Hurricanes Katrina and Rita (Category 3), many cooling towers sustained component and cladding damage and loss of fan shrouds (Figures 4-13 and 4-14).

4.2.4 Components and Cladding

Attachment of components and cladding is a very important consideration during the design of a structure. Figures 4-15 and 4-16 depict the typical type of cladding failure on buildings caused by high wind velocities and the importance of considering the corner attachment as in Figure 4-15.

Figure 4-8. Power pole, Hurricane Frances.
Source: Courtesy of NASA KSC.

Figure 4-9. Cable tray damage, Hurricane Frances.
Source: Courtesy of NASA KSC.

Figure 4-10. Utility pole, Hurricane Rita.
Source: Courtesy of FEMA.

Figure 4-11. Power poles, Hurricane Rita.
Source: Courtesy of NIST.

Figure 4-12. Electrical transmission tower, Hurricane Rita.
Source: Courtesy of NIST.

Figure 4-13. Cooling tower damage, Hurricane Katrina.
Source: Courtesy of LCA.

Figure 4-14. Cooling tower damage, Hurricane Katrina.
Source: Courtesy of LCA.

Figure 4-15. Cladding damage, Hurricane Frances.
Source: Courtesy of NASA KSC.

Figure 4-16. Tank cladding damage, Hurricane Rita.
Source: Courtesy of LCA.

Figure 4-17. Example of minor wind-related damage to temporary facilities and construction supply storage tents in Gregory, Texas, Hurricane Harvey.

Figure 4-18. Flare tower damage, Hurricane Rita.
Source: Courtesy of FEMA.

During Hurricanes Katrina and Rita (Category 3), tank cladding damage was reported.

Cladding on temporary structures can also be vulnerable to wind damage as shown in Figure 4-17.

4.2.5 Tall Vertical Equipment

Careful consideration should be given to secure, demount, or disassemble tall vertical equipment, such as flare towers (Figure 4-18) and tower cranes (Figure 4-19) during a high-wind event to ensure the components providing the lateral support are not compromised and result in failure.

4.3 WIND TUNNEL TESTING

ASCE 7 allows the use of wind tunnel testing to estimate wind loads in lieu of the simplified and analytical procedures. The recommended provisions in this guide are classified with the analytical procedure for "other structures." The use of the analytical procedure is limited, and ASCE 7, Section 26.1.2.1 requires that structures be designed using either recognized literature documenting the load effects or wind tunnel testing in cases where

Figure 4-19. Crane boom across pipe back, Hurricane Rita.
Source: Courtesy of LCA.

- A structure may have across-wind loading,
- Vortex shedding may occur,
- Instabilities due to galloping or flutter are possible,
- A structure has a site location for which channeling effects or buffeting in the wake of upwind structures warrant special consideration,
- A structure has unusual shape, or
- A structure has unusual response characteristics.

In addition to these limitations on analytical procedures, there are other considerations particular to petrochemical structures that may make wind tunnel testing an attractive alternative method:

- A structure is not adequately covered by the provisions of this book. This limitation is similar to the limitation regarding unusually shaped structures.
- It is suspected that the currently available wind load analysis methods may be overly conservative, and a more precise understanding of wind loads may result in a more efficient structure. For example, the analytical procedure explicitly prohibits reductions in velocity pressure due to apparent shielding from neighboring structures.

- The adequacy of an existing structure to support new or modified equipment or appurtenances is being investigated.
- A validation of wind load analysis practices is desired.

If a wind tunnel test is to be used to estimate wind loads or wind load response for a structure, it is important that the simulation properly represents the velocity profile and turbulence characteristics of the atmospheric boundary layer; the structures and surrounding topography are geometrically similar to the prototype; wind tunnel blockage is limited; wind tunnel pressure gradients are accounted for; Reynolds number effects are minimized; and the response characteristics of the instrumentation are consistent with the planned measurements. These requirements are listed in ASCE 7-16, Section 31.2, and more detailed guidance on proper wind tunnel testing practice is contained in ASCE 49-12 (2012). Particular care should be taken in minimizing Reynolds number effects. Because of the presence of exposed framing elements, equipment, and appurtenances that are common with petrochemical structures, length scales far below what is typical for enclosed buildings will be present. This complicates the determination of characteristic length for dynamic similarity considerations.

Loads for the main wind-force-resisting system determined by wind tunnel testing shall meet the limitations stated in ASCE 7-16, Section 31.4.4 and Commentary C31.

There are three common model types used in wind tunnel testing of civil engineering structures: rigid pressure models, force balance models, and aeroelastic models. Rigid pressure models are used to measure surface pressures on building models. Small holes on the surfaces of these models transmit surface pressures to pressure transducers via thin tubing. Pressure model tests are particularly useful for determining the peak pressure coefficients that control the design of components and cladding with small tributary areas. As the price of electronic instrumentation falls, simultaneous measurement and integration of surface pressures is becoming a more common technique for estimating overall building loads.

Force balance models are typically used for determining the overall force, moment, and torsion coefficients for a particular structure. A force balance is an instrument consisting of load cells and/or strain gauges. When the fundamental frequency of the model/force balance system is high, this technique can be used to measure the overall fluctuating forces and moments generated by the interaction of the turbulence in the approaching wind and structure geometry. The fluctuating forces can be used in conjunction with the structure's response mode shapes and frequencies to estimate the dynamic response of the prototype in natural winds.

Aeroelastic models are used for structures that are so flexible that there is a significant coupling between the aerodynamic forces and the dynamic response of the structure. A proper aeroelastic model correctly represents the mass, stiffness, and damping characteristics of the prototype structure at model scale. These are the most expensive and complicated wind tunnel tests and are commonly used for structures such as long-span bridges.

Wind tunnel testing can be expensive and time-consuming, and therefore is not typically used for the design of new structures. However, when evaluating existing facilities or critical applications, wind tunnel testing provides reliable predictions of wind forces when performed correctly.

4.4 CFD COMMENTARY

The determination of wind loads on many types of structures including some of the structures that make up a petrochemical facility is based, in large part, on the results of wind tunnel testing. Such tests are conducted either on full-scale component members or on scale models of actual structural configurations. Although wind tunnel testing is a recognized and well-established method of determining wind loads, in select cases CFD modeling may be a suitable alternative method for evaluating wind loads on structures. In addition, CFD modeling can often be employed in conjunction with wind tunnel simulations to streamline the design process while still returning to the wind tunnel for final validation.

CFD is a technique that uses numerical models to compute fluid flow around or through a body. When applied properly, the benefits are obvious. By understanding the effects of wind flow acting on a structure early in the design process, designers can alter the structural configuration to optimize performance. For this reason, the automotive industry has invested heavily in CFD models to accelerate the development of new vehicle models by applying CFD models early in the design process.

Although modeling fluid flow is an evolving science, CFD models have advanced considerably since their introduction in the 1970s. Most CFD models require solving a highly nonlinear set of partial differential equations known as Navier–Stokes equations. Used to describe fluid motion mathematically, the Navier–Stokes equations provide a relationship between density, pressure, viscosity, and velocity that conserves mass and momentum. However, they cannot be solved exactly except for special cases. Hence, the standard approach to obtain a solution is to divide the flow domain into a large number of small cells, then calculate the fluid parameters on a rigid grid. For more complex problems, this approach can cause physical and numerical instabilities that in turn can cause calculations to abort. Because these equations sometimes experience numerical instabilities, sophisticated numerical techniques are required to achieve a stable solution that reasonably matches experimental results.

CFD models using Navier–Stokes equations assume that the fluid is a continuum material, rather than a collection of molecules. Another CFD approach is to use what physicists refer to as the "lattice Boltzmann method" in place of the Navier–Stokes equations. A CFD model based on the lattice Boltzmann method is basically a discrete particle system having a finite number of states that are updated in distinct steps. Each step follows either deterministic or stochastic rules dictated by the conservation of mass, momentum, and energy. The resulting behavior is equivalent to, but independent of, continuum mechanics. The solution produced by the lattice Boltzmann method is identical to the solution produced by Navier–Stokes equations, yet no differential equations are being solved by the process.

The flow around structures will typically be very turbulent. Turbulence in CFD models can be resolved in several different ways. Reynolds-averaged Navier-Stokes equations will divide the flow between a mean (averaged) portion and a fluctuating portion. Various closure methods are available to solve for the fluctuating term. This method will produce a time-averaged field that does not resolve all turbulent scales. Calculation methods that resolve more of the fluctuating fields are available but may be unsuitable for some applications. Direct numerical simulation will simulate flow over all time and length scales but can result in excessive computational burdens. Large-eddy simulations will resolve only the larger-scale variations while filtering the more computationally intensive small-scale fluctuations.

All CFD models have had mixed results in reproducing the separation and reattachment of fluids at high wind velocities when the Reynolds number is large and the flow is turbulent. Estimations of mean pressures in zones of separation and reattachment are hence more likely to be suspect, while errors in predicting mean and fluctuating surface pressures may also lead to errors in predicting overall wind loads for certain geometries. Another challenge from a computational standpoint is the complexity of a three-dimensional geometry associated with structures like an open-frame process tower that is made up of a collection of round and sharp-edged members. However, rapid advancements in computer designs and operating system development have made this challenge less daunting.

When determining wind loads on structures that make up a petrochemical facility, cases may arise that justify the use of a CFD model. In such cases, careful consideration should be given to ensuring the CFD model is properly applied during analysis of the structure or component. When using a commercially available CFD program, request that the vendor provide case studies of the application of their model to wind-related analyses, including benchmarking to wind tunnel or full-scale measurements. Also, be mindful of the computing requirements to execute the CFD runs. For complex geometries, very fine grid resolution may be required. CFD simulations should be tested for grid independence to demonstrate that finer resolution would not substantially change the final conclusions. Finally, recognize that some complex geometries involving a multitude of assemblies may prove too difficult to solve in practice because of computing hardware limitations.

4.5 LOAD COMBINATIONS

This section addresses load combinations that include the effects of wind loading. Other load combinations without wind loading are not considered in this guideline.

The engineer is responsible for verifying applicable codes and standards by state or local agencies that have jurisdiction in the state the project is constructed.

4.5.1 References to Existing Industry Guidelines

Building codes, standards, and industry guidelines address load combinations for buildings and structures. This section highlights some of the main industry standards used for loading and load combinations; it is not meant to be an exhaustive list. The following references discuss load combinations in the presence of wind loading.

4.5.1.1 ASCE 7-16 and International Building Code 2018

The basic load combinations set forth in Chapter 2 of ASCE 7 and Chapter 16 of the International Building Code (IBC) (ICC 2018) are the minimum to be considered for the design of buildings and structures. While these codes are primarily geared to buildings, the load combinations define the basic concepts for addressing the effects of wind loads in combination with other loads.

4.5.1.2 Process Industry Practices STC01015

Because ASCE 7 and IBC do not address loads and load combinations specifically for petrochemical or industrial facilities, Process Industry Practices (PIP) STC01015 was developed to assist engineers in developing consistent practices for the design of these facilities.

In general, PIP STC01015, Section 4.2 is based on the combinations set forth in ASCE 7 and IBC. However, loads and load combinations for particular petrochemical and industrial structures and equipment are also addressed. Tables 4-1 and 4-2 provide the STC01015 load combinations for buildings and open-frame structures. Variations on these combinations are also provided in STC01015 for vertical vessel supports and foundations, horizontal vessel heat exchanger supports and foundations, pipe racks and pipe bridges, and ground-supported storage tank foundations.

PIP STC01015 no longer deviates from ASCE 7 for uplift and overturning load cases. Both the factor of safety for overturning and sliding for load combinations with wind shall be 1.0. As noted in PIP STC01015, the "factor of safety" for overturning and sliding is built into the load factors in the load combinations addressing these conditions.

Owners' and local building authorities' requirements should be consulted and may be different than those discussed in this guideline.

Tables 4-1 and 4-2 are examples of the load combinations from PIP STC01015. Again, these are only the combinations designated for buildings and

Table 4-1. IBC/ASCE Load Combinations That Include Wind with PIP Equivalent: Strength Design Load Combinations for Buildings and Open-Frame Structures.

IBC/ASCE 7 load combination	*PIP STC01015 load combination*	*PIP load combination*	*Description*
$1.2D + 1.6(L_r$ or S or $R) + 0.5W$	5-3.2	$1.2(D_s + D_o) + 1.2T_s + 1.6(L_r$ or S or $R) + 0.5W$	Operating weight + sustained thermal + (roof live or snow or rain) + wind
	5-3.4	$1.2(D_s + D_t) + 1.2T_s + 1.6(L_r$ or S or $R) + 0.5W$	Test weight + sustained thermal + (roof live or snow or rain) + wind
$1.2D + 1.0W + L + 0.5(L_r$ or S or $R)$	5-4.1	$1.2(D_s + D_o) + 1.2T_s + 1.0W + 0.5L + 0.5(L_r$ or S or $R)$	Operating weight + sustained thermal + wind + live + (roof live or snow or rain)
	5-4.2	$1.2(D_s + D_t) + 1.2T_s + 1.0W_p + 0.5L + 0.5(L_r$ or S or $R)$	Test weight + sustained thermal + partial wind + live + (roof live or snow or rain)
$0.9D + 1.0W$	5-6.1	$0.9(D_s + D_e) + 1.2T_s + 1.0W$	Empty weight + sustained thermal + wind (wind uplift case)
	5-6.2	$0.9(D_s + D_o) + 1.2T_s + 1.0W$	Operating weight + sustained thermal + wind (sustained thermal and wind uplift case)

Table 4-2. IBC/ASCE Load Combinations That Include Wind with PIP Equivalent: Allowable Stress Design Load Combinations for Buildings and Open-Frame Structures.

IBC/ASCE 7 load combination	*PIP STC01015 load combination*	*PIP load combination*	*Description*
$D + 0.6W$	4-5.1	$D_s + D_o + T_s + 0.6W$	Operating weight + sustained thermal + wind
	4-5.3	$D_s + D_t + T_s + 0.6W_p$	Test weight + sustained thermal + partial wind
$D + 0.75L + 0.75(0.6W) + 0.75(L_r$ or S or $R)$	4-6a.1	$D_s + D_o + T_s + 0.75L + 0.75(0.6W) + 0.75(L_r$ or S or $R)$	Operating weight + sustained thermal + live + wind + (roof live or snow or rain)
	4-6a.2	$D_s + D_t + T_s + 0.75L + 0.75(0.6W_p) + 0.75(L_r$ or S or $R)$	Test weight + sustained thermal + live + wind + (roof live or snow or rain)
$0.6D + 0.6W$	4-7.1	$0.6(D_s + D_e) + T_s + 0.6W$	Empty weight + sustained thermal + wind (sustained thermal + wind uplift case)
	4-7.2	$0.6(D_s + D_o) + T_s + 0.6W$	Operating weight + sustained thermal + wind (sustained thermal and wind uplift case)

open-frame structures that include the wind variable. There are a number of other load combinations that must be considered. Similar concepts can be applied to the tables for other types of structures in PIP STC01015. The tables in PIP STC01015 have several important footnotes to help the user correctly apply the load combinations that should be reviewed carefully.

4.5.2 Supplemental Guidelines for Petrochemical and Other Industrial Facilities

This guideline supplements the guidance of PIP STC01015 and provides additional guidance regarding load cases including wind for consistent application for new structures and the evaluation of existing structures.

Load combinations for new structures need to take into consideration the full-time history of the facility, from construction through long-term operation. Load conditions such as construction, empty, hydrostatic test, and upset condition that are common for petrochemical facilities are not included in ASCE 7 and are addressed in the following.

4.5.2.1 Load Definitions

Loads for petrochemical facilities are unique in the way they vary over time. For instance, dead load can be fairly complex in that a specific piece of equipment may have four to five weight conditions, depending on the stage of construction or operating condition.

Similarly, wind loads can vary over different loading conditions. For new facilities, under normal operating conditions, maximum wind loads should be utilized. However, for temporary conditions, such as construction and hydrostatic test, a reduced wind load may be considered. See ASCE 37-14 (ASCE 2014) for guidance.

The following load definitions are to be considered for use:

- D_s = Dead load due to self-weight of structure, including foundation weight, soil overburden, and fireproofing, as appropriate for the load condition being considered. Fireproofing is to be considered applicable for the structure and load condition considered. Field-applied fireproofing should not be considered for the construction load condition, whereas shop-applied fireproofing would be applicable in all load conditions.
- D_e = Dead load due to the empty weight of piping, electrical trays, and equipment. Many times, the weight of piping associated with a vessel is estimated as a percentage of the empty weight of a vessel. Caution must be exercised not to overestimate these loads for the empty or uplift conditions.
- D_f = Dead load due to fabricated weight of equipment, prior to appurtenances such as platforms, ladders, and internals that are added during construction.
- D_o = Dead load due to operating weight of piping, electrical trays, and equipment (including fluid).

- D_t = Dead load due to test weight of piping and equipment.
- D_u = Load due to upset load conditions for vessels during operation. Upset loads typically exceed normal operating loads. Additionally, loads from upset conditions are commonly from process surge forces that may be considered as a live load for new construction.
- D_h = Dead load of equipment per the facility's hurricane procedures to be considered for evaluation of existing structures.
- L = Live loads from platforms, equipment surge loads, and impact loads.
- L_r = Roof live load.
- R = Rain load.
- S = Snow load per ASCE 7, appropriate for the project location.
- T_s = Sustained thermal (and/or self-straining) loads owing to pipe anchor or guide loads, exclusive of friction loads.
- T_t = Temporary thermal (and/or self-straining) loads due to pipe anchor or guide loads, exclusive of friction loads. Only thermal loads expected during the temporary period need to be considered.
- W = Wind load per ASCE 7, appropriate for the project location.
- W_{NH} = Wind load based on non-hurricane basic wind speed (applicable to hurricane-prone regions only). Engineer may consider using wind speeds from ASCE 37-14, which uses a wind speed of 115 mi/h for Risk Category II and 122 mi/h for Risk Category III as the non-hurricane basic wind speed for structures. Alternatively, the engineer could use a site-specific meteorological wind study to determine the appropriate non-hurricane wind speeds.
- W_p = Partial wind load for use in short-term conditions such as hydrostatic testing of equipment, construction, or turnaround.

Construction Load Cases

Load combinations for construction must consider the different stages of construction including the varying completion of the structure. Overturning during this time period may be a controlling design scenario because the equipment may be empty or incomplete (D_f), and therefore the total weight resisting overturning would be less. However, the lesser weight may be paired with a lower wind speed to minimize the impact. Per PIP STC01015 and ASCE 37 the partial wind speed to be used in PIP STC01015 (W_p) for construction can be calculated as follows:

- The partial wind load should be the ASCE 7 design wind speed, adjusted by the appropriate factor from ASCE 37, dependent on the estimated duration of construction or other activity. These factors range from 0.75 (for a 6-week duration) to 0.9 if duration is anticipated to be 2 to 5 years. If the construction is known to occur outside of hurricane season, then a basic reduced wind speed may be used.

- If the construction is known to occur during hurricane season, then the reduced basic wind speed may be used with the appropriate factor from ASCE 37, provided additional bracing is prepared in advance and will be applied prior to hurricane landfall.
- If construction may occur during hurricane season and additional bracing is not prepared in advance, the full basic wind speed from ASCE 7 shall be used.
- Lower wind speeds may be used per ASCE 37 for continuously monitored work periods.

Caution: Even when construction is estimated to occur outside of hurricane season, the engineer should consider bracing provisions as a potential mitigation for schedule slippage that may shift construction into the hurricane weather window.

For short-term conditions, use of Risk Category II may be considered because the consequence of failure during construction or testing may be different than a failure during normal operation.

Empty Load Cases

Similar to the construction load cases, an empty load case should be considered for overturning. In this scenario, the structure and equipment construction are complete but may be empty (D_e). Because this scenario may occur at any time, the full wind load case should be used for design of new structures.

Hydrostatic Test Load Cases

For load combinations considering hydrostatic test loads (D_t), a reduced wind load may be used similar to that described for the construction load case in Table 4-5. In addition, in lieu of additional bracing if the hydrostatic test can be deferred or stopped in advance of an approaching hurricane, a reduced wind speed may be used with the appropriate factor from ASCE 37 in combination with the hydrostatic test loads (D_t), while the full design wind load would be used with a dead load specific to the shutdown procedure (D_h), which typically would not control over the operating condition used for design of new structures.

For extremely short test periods (days), consideration can be made to place a limit on environmental conditions when the test may be performed. Doing so may mean testing delay and increasing construction costs and schedule and should be done with caution.

In addition, for test conditions, use of Risk Category 2 may be considered because the consequence of failure during testing may be different than a failure during normal operation.

Upset Load Cases

It is the opinion of this committee that the upset load should be considered an infrequent load condition, such that the full hurricane-level wind load and full upset load need not be considered simultaneously. However, it is recommended to consider the upset load in combination with a non-hurricane wind load (W_{NH}).

Table 4-3. Supplemental Strength Design Load Combinations That Include Wind for PIP STC01015 for New Structures.

IBC/ASCE 7 load combination	*PIP STC01015 load combination*	*PIP load combination*	*Description*
$1.2D + 1.6(L_r \text{ or } S \text{ or } R) + 0.5W$	5-3.2	$1.2(D_s + D_o) + 1.2T_s + 1.6(L_r \text{ or } S \text{ or } R) + 0.5W$	Operating weight + sustained thermal + (roof live or snow or rain) + wind
	—	$1.2(D_s + D_t) + 1.2T_s + 1.6(L_r \text{ or } S \text{ or } R) + 0.5W_p$	Test weight + sustained thermal[a] + (roof live or snow or rain) + partial wind
	—	$1.2(D_s + D_f) + 1.2T_t + 1.6(L_r \text{ or } S \text{ or } R) + 0.5W_p$	**Construction weight + construction thermal + (roof live or snow or rain) + partial wind**
$1.2D + 1.0W + L + 0.5(L_r \text{ or } S \text{ or } R)$	5-4.1	$1.2(D_s + D_o) + 1.2T_s + 1.0W + 0.5L + 0.5(L_r \text{ or } S \text{ or } R)$	Operating weight + sustained thermal + wind + live + (roof live or snow or rain)
	5-4.2	$1.2(D_s + D_t) + 1.2T_s + 1.0W_p + 0.5L + 0.5(L_r \text{ or } S \text{ or } R)$	Test weight + sustained thermal + wind + live + (roof live or snow or rain)
	—	$1.2(D_s + D_f) + 1.2T_s + 1.0W_p + 0.5L + 0.5(L_r \text{ or } S \text{ or } R)$	**Construction weight + sustained thermal[a] + partial wind + live + (roof live or snow or rain)**
	—	$1.2D_s + 1.2T_s + 1.0W_{NH} + 1.0D_u$	Weight + sustained thermal + nonhurricane wind + upset weight[b]
$0.9D + 1.0W$	5-6.1	$0.9(D_s + D_e) + 1.2T_s + 1.0W$	Empty weight + sustained thermal + wind (wind uplift case)
	5-6.2	$0.9(D_s + D_o) + 1.2T_s + 1.0W$	Operating weight + sustained thermal + wind (sustained thermal and wind uplift case)
	—	$0.9(D_s + D_f) + 1.2T_t + 1.0W_p$	**Construction weight + construction thermal + partial wind (wind uplift case)**

Notes:

1. Bold text notes load combinations that have been revised from PIP STC01015.
2. If the live load is greater than 100 psf, then the live load factor shall be 1.0.

[a]Only T_s loads expected during hydrostatic testing period need to be considered.

[b]Additional loads for upset conditions are commonly from process surge forces, which may be considered as a live load for new construction.

Table 4-4. Supplemental Allowable Stress Design Load Combinations That Include Wind for PIP STC01015 for New Structures.

IBC/ASCE 7 load combination	*PIP STC01015 load combination*	*PIP load combination*	*Description*
$D + 0.6W$	4-5.1	$D_s + D_o + T_s + 0.6W$	Operating weight + sustained thermal + wind
	4-5.3	$D_s + D_t + T_s + 0.6W_p$	Test weight + sustained thermal[a] + partial wind
	—	$D_s + D_f + T_t + 0.6W_p$	**Construction weight + construction thermal + partial wind**
$D + 0.75L + 0.75(0.6W) + 0.75(Lr \text{ or } S \text{ or } R)$	4-6a.1	$D_s + D_o + T_s + 0.75L + 0.75(0.6W) + 0.75(L_r \text{ or } S \text{ or } R)$	Operating weight + sustained thermal + live + wind + (roof live or snow or rain)
	4-6a.2	$D_s + D_t + T_s + 0.75L + 0.75(0.6W_p) + 0.75(L_r \text{ or } S \text{ or } R)$	Test weight + sustained thermal[a] + live + wind + (roof live or snow or rain)
	—	$D_s + D_f + T_t + 0.75L + 0.75(0.6W_p) + 0.75(L_r \text{ or } S \text{ or } R)$	**Construction weight + construction thermal + live + wind + (roof live or snow or rain)**
	—	$D_s + T_s + 0.75D_u + 0.75(0.6W_{NH})$	**Weight + sustained thermal + upset weight[b] + nonhurricane wind**
$0.6D + 0.6W$	4-7.1	$0.6(D_s + D_e) + T_s + 0.6W$	Empty weight + sustained thermal + wind (wind uplift case)
	4-7.2	$0.6(D_s + D_o) + T_s + 0.6W$	Operating weight + sustained thermal + wind (wind uplift case)
	—	$0.6(D_s + D_f) + T_t + 0.6W_p$	**Construction weight + construction thermal + partial wind**

Notes:

1. Bold text notes load combinations that have been revised from PIP STC01015.

[a]Only T_s loads expected during hydrostatic testing period need to be considered.

[b]Additional load for upset conditions are commonly from process surge forces, which may be considered as a live load for new construction.

4.5.2.2 Load Combinations for the Design of New Structures

Tables 4-3 and 4-4 provide additional and modified wind load combinations to supplement those provided by PIP STC01015. The tables have been expanded to address construction and upset load combinations.

4.5.2.3 Load Combination Considerations for Existing Structures

General

Modification to existing structures is often a balance between the reduction of risk and the cost of modification. In an effort to maximize the benefit and minimize the cost of an upgrade, most facility owners are receptive to practical design considerations. The engineer should contact the owner in the early stages of design to make recommendations and agree on special load combination provisions for the evaluation of existing structures. Once alignment with the owner is achieved in the areas discussed subsequently, consultation and agreement from the local building officials where the facility falls under a local jurisdiction may be required. The term *existing structures* is intended to include process structures, shelters, pipe racks, and equipment. The purpose of these recommendations is to provide the engineer and owner with options in developing the appropriate level and cost of modifications.

Hurricane Event Considerations

In new structure design, hurricane wind speeds are utilized for normal operating conditions, regardless of the facility's hurricane response plan. Unlike seismic events, hurricane events are known in advance and the facility typically has a hurricane response plan that includes shutdown procedures. Facilities are not normally left operating during a hurricane and the engineer should consult with the owner for their site-specific hurricane response procedures. If well-documented hurricane shutdown procedures exist, the following items should be considered for the evaluation and modification of existing structures with the owner's agreement.

Operational Gravity Loads

Based on the facility hurricane procedures, vessels and tanks may be left with fluids inside to resist overturning forces from a hurricane event, while some may have inventories reduced. Depending on inventories and chemistry, tanks may be filled with water. The owner should be consulted on anticipated weights for major equipment and items during hurricane conditions based on the hurricane shutdown plan. For evaluation of existing structures under hurricane wind speeds, the appropriate dead load (D_h) of equipment per the facility's hurricane procedures should be utilized.

Table 4-5. Strength Design Load Combinations That Include Wind for PIP STC01015 for Evaluation of Existing Structures.

IBC/ASCE 7 load combination	*PIP STC01015 load combination*	*PIP load combination*	*Description*
$1.2D + 1.6(L_r \text{ or } S \text{ or } R) + 0.5W$	5-3.2	$1.2(D_s + D_o) + 1.2T_s + 1.6(L_r \text{ or } S \text{ or } R) + 0.5W_{NH}$	Operating weight + sustained thermal + (roof live or snow or rain + **nonhurricane wind**
	5-3.2	$1.2(D_s + D_t) + 1.2T_s + 1.6(L_r \text{ or } S \text{ or } R) + 0.5W_p$	Test weight + sustained thermal[a] + (roof live or snow or rain) + partial wind
	—	$1.2(D_s + D_f) + 1.2T_t + 1.6(L_r \text{ or } S \text{ or } R) + 0.5W_p$	**Construction weight + construction thermal + (roof live or snow or rain) + partial wind**
	—	$1.2(D_s + D_h) + 1.2T_t + 1.6(S \text{ or } R) + 0.5W$	**Shutdown weight + sustained thermal[b] + (snow or rain)**
$1.2D + 1.0W + L + 0.5(L_r \text{ or } S \text{ or } R)$	5-4.2	$1.2(D_s + D_o) + 1.2T_s + 1.0W_{NH} + 0.5L + 0.5(L_r \text{ or } S \text{ or } R)$	Operating weight + sustained thermal + **nonhurricane wind** + live[c] + (roof live or snow or rain)
	5-4.2	$1.2(D_s + D_t) + 1.2T_s + 1.0W_p + 0.5L + 0.5(L_r \text{ or } S \text{ or } R)$	Test weight + sustained thermal + partial wind + live[c] + (roof live or snow or rain)
	—	$1.2(D_s + D_f) + 1.2T_s + 1.0W_p + 0.5L + 0.5(L_r \text{ or } S \text{ or } R)$	**Construction weight + sustained thermal[a] + partial wind + live[c] + (roof live or snow or rain)**
	—	$1.2D_s + 1.2T_s + 1.0W_{NH} + 1.0D_u$	**Weight + sustained thermal + nonhurricane wind + upset weight [d]**
	—	$1.2(D_s + D_h) + 1.2T_t + 1.0W + L_h + 0.5(L_r \text{ or } S \text{ or } R)$	**Shutdown weight + sustained thermal[b] + wind + live[e] + (roof live or snow or rain)**

$0.9D + 1.0W$	5-6.1	$0.9(D_s + D_e) + 1.2T_s + 1.0W_{NH}$	Empty weight + sustained thermal + **wind** (wind uplift case)
	5-6.2	$0.9(D_s + D_o) + 1.2T_s + 1.0W_{NH}$	Operating weight + sustained thermal + **wind** (wind uplift case)
	—	$0.9(D_s + D_f) + 1.2T_t + 1.0W_p$	**Construction weight – construction thermal + construction wind (wind uplift case)**
	—	$0.9(D_s + D_h) + 1.2T_t + 1.0W$	**Shutdown weight + sustained thermal**[b] **+ wind (wind uplift case)**

Notes:

1. Bold text notes load combinations that have been revised from PIP STC01015.

[a]Only T_s loads expected during hydrostatic testing period need to be considered.

[b]Only T_t loads expected during shutdown period need to be considered.

[c]If the live load is greater than 100 psf, then the live load factor shall be 1.0.

[d]Additional loads for upset conditions are commonly from process surge forces, which may be considered as a live load for evaluation.

[e]A reduced live load based on the actual expected conditions during a hurricane may be considered.

Table 4-6. Allowable Stress Design Load Combinations That Include Wind for PIP STC01015 for Evaluation of Existing Structures.

IBC/ASCE 7 load combination	*PIP STC01015 load combination*	*PIP load combination*	*Description*
$D + 0.6W$	—	$D_s + D_o + T_s + 0.6W_{NH}$	Operating weight + sustained thermal + **nonhurricane wind**
	4-5.3	$D_s + D_t + T_s + 0.6W_p$	Test weight + sustained thermal[a] + partial wind
	—	$D_s + D_f + T_t + 0.6W_p$	**Construction weight + construction thermal + partial wind**
	—	$D_s + D_h + T_t + 0.6W$	**Shutdown weight + sustained thermal[b] + wind**
$D + 0.75L + 0.75(0.6W) + 0.75(L_r \text{ or } S \text{ or } R)$	4-6a.1	$D_s + D_o + T_s + 0.75L + 0.75(0.6W_{NH}) + 0.75(L_r \text{ or } S \text{ or } R)$	Operating weight + sustained thermal + live + **nonhurricane wind** + (roof live or snow or rain)
	4-6a.2	$D_s + D_t + T_s + 0.75L + 0.75(0.6W_p) + 0.75(L_r \text{ or } S \text{ or } R)$	Test weight + sustained thermal[a] + live + partial wind + (roof live or snow or rain)
	—	$D_s + D_f + T_t + 0.75L + 0.75(0.6W_p) + 0.75(L_r \text{ or } S \text{ or } R)$	**Construction weight + construction thermal + live + partial wind + (roof live or snow or rain)**
	—	$D_s + T_s + 0.75D_u + 0.75(0.6W_{NH})$	**Weight + sustained thermal + upset weight[c] + nonhurricane wind**
	—	$D_s + D_h + T_s + L + 0.75(0.6W) + 0.75(L_r \text{ or } S \text{ or } R)$	**Shutdown weight + sustained thermal[b] + live[d] + wind + (roof live or snow or rain)**

$0.6D + 0.6W$**(For structure design and stability check)** [e]	4-7.1	$0.6(D_s + D_e) + T_s + 0.6W_{NH}$	Empty weight + sustained thermal + **nonhurricane wind** (wind uplift case)
	4-7.2	$0.6(D_s + D_o) + T_s + 0.6W_{NH}$	Operating weight + sustained thermal + **nonhurricane wind** (wind uplift case)
	—	$0.6(D_s + D_f) + T_t + 0.6W_p$	**Construction weight + construction thermal + partial wind**
	—	$0.6(D_s + D_h) + T_s + 0.6W$	**Shutdown weight + sustained thermal[b] + wind**
$0.6D + 0.6W$**(For foundation design and pile tension check)** [e,,f]	—	$0.9(D_s + D_e) + T_s + 0.6W_{NH}$	**Empty weight + sustained thermal + nonhurricane wind (wind uplift case)**
	—	$0.9(D_s + D_o) + T_s + 0.6W_{NH}$	**Operating weight + sustained thermal + nonhurricane wind (wind uplift case)**
	—	$0.9(D_s + D_f) + T_t + 0.6W_p$	**Construction weight + construction thermal + partial wind**
	—	$0.9(D_s + D_h) + T_s + 0.6W$	**Shutdown weight + sustained thermal[b] + wind**

Notes: Bold text notes load combinations that have been revised from PIP STC01015.
[a]Only T_s loads expected during hydrostatic testing period need to be considered.
[b]Only T_t loads expected during shutdown period need to be considered.
[c]Additional load for upset conditions are commonly from process surge forces, which may be considered as a live load for evaluation.
[d]A reduced live load based on the actual expected conditions during a hurricane may be considered.
[e]When performing stability and foundation design, a stability ratio (against overturning and sliding) greater than or equal to 1.0 should be provided. Effects of buoyancy should be considered. Factors of safety for stability ratio are built into ASCE 7 load combinations.
[f]It is the opinion of this committee that for evaluation of existing structures, the same level of conservatism in the calculation of bearing pressure and pile tension may not be warranted. The committee recommends the use of 0.9*D* while developing calculations for bearing pressure and pile tension checks because the factor of safety for soil capacity is typically included in the geotechnical report recommendations.

Live Loads

During hurricane events where the facility is in shutdown mode, nonessential personnel are evacuated, and only essential staff remains. In this case, live loads would be minimal or nonexistent and may therefore be reflected in the load combinations. Typically, the highest live loads are during construction, turnaround, and maintenance activities because of personnel and temporary loads. In some instances, operational conditions, such as catalyst staging areas, create high localized live loads. However, these activities would be performed in minimal wind conditions, significantly less than used for design purposes. Therefore, reduced live load should be considered for normal wind load combinations for modification to existing structures.

Friction Loads

Unless otherwise defined by hurricane shutdown procedures, because the plant will be in shutdown mode, friction loads should not be considered during hurricane conditions. As appropriate, cold spring and/or anchor loads may need to be considered. Any special hurricane load or load combination design basis and assumptions utilized for the evaluation of existing structures should be clearly documented on structural and process drawings and clearly indicated in the hurricane response plan and the facility shutdown procedures.

Load Combinations for the Evaluation of Existing Structures

The recommended modifications to the PIP STC01015 load combinations for evaluation of existing structures are provided in Tables 4-5 and 4-6.

4.6 WIND LOADS FOR LNG FACILITIES

4.6.1 Introduction

Natural gas continues to serve an essential role in the US energy supply and is quickly becoming a popular fuel resource both domestically and internationally. Globally, it is extensively used as a fuel to meet the needs of residential, commercial, and industrial purposes. To meet this growing demand, there has been increased activity within the last decade or so related to new development and construction of LNG facilities within the United States. Depending on their location and intended commercial use, LNG facilities are capable of liquefying and exporting domestically produced natural gas and/or importing and regasifying foreign-sourced LNG. These facilities consist of equipment and structures similar to those typically found in petrochemical or industrial facilities. In the United States, several federal agencies have regulatory authority over LNG facilities. They include the Federal Energy Regulatory Commission (FERC), the US Coast Guard, and the US DOT Pipeline and Hazardous Materials Safety Administration (PHMSA). The main intent of this section is to provide more consistent wind load design guidance for LNG facilities using ASCE 7 to meet the requirements of the *Code of Federal Regulations.*

Wind load design for LNG facilities used in conjunction with the transportation of gas by pipeline in the United States is governed by PHMSA pursuant to 49 CFR §193.2067. 49 CFR §193.2001(a) clarifies that the requirements in Part 193 apply to "LNG facilities used in the transportation of gas by pipeline that is subject to the pipeline safety laws (49 U.S.C. 60101 *et seq.*) and Part 192 of this chapter." 49 CFR §193.2067(b) references ASCE 7-05 (ASCE 2005) (incorporated by reference in 49 CFR §193.2013) for shop-fabricated containers and specifies a prescriptive wind speed for all other LNG facilities. The LNG facility operator must comply with all applicable federal safety regulations, including 49 CFR Part 193, as well as verifying which industry codes and standards apply for the specific facility location based on relevant state and local building authority jurisdiction requirements. In the event of a conflict between 49 CFR Part 193 and state and local code, 49 CFR Part 193 prevails. For locations outside of the United States, local code requirements are to be satisfied at a minimum.

As part of the permitting process for FERC jurisdictional LNG facilities, the LNG facility operator must submit the wind load design basis for Part 193, Subpart B compliance review by PHMSA and an engineering and technical review and approval by FERC to comply with each respective agency's statutory authority. For all other PHMSA jurisdictional LNG facilities, the LNG facility operator must demonstrate compliance with Part 193 through the PHMSA inspection program.

4.6.2 Code of Federal Regulations

Requirements of 49 CFR §193.2067 (at the time of publication of this document) are as follows:

§193.2067 Wind forces.

(a) LNG facilities must be designed to withstand without loss of structural or functional integrity;

1. The direct effect of wind forces;
2. The pressure differential between the interior and exterior of confining, or partially confining, structure; and
3. In the case of impounding systems for LNG storage tanks, impact forces and potential penetrations by wind borne missiles.

(b) The wind forces at the location of the specific facility must be based on one of the following:

1. For shop fabricated containers of LNG or other hazardous fluids with a capacity of not more than 70,000 gallons, applicable wind load data in ASCE/SEI 7-05.

2. For all other LNG facilities:

(i) An assumed sustained wind velocity of not less than 150 miles per hour, unless the Administrator finds a lower velocity is justified by adequate supportive data; or

(ii) The most critical combination of wind velocity and duration, with respect to the effect on the structure, having a probability of exceedance in a 50-year period of 0.5 percent or less, if adequate wind data are available and the probabilistic methodology is reliable.

This guidance provides an interpretation of Part 193.2067(b)(2), and "all other LNG facilities." The guidance herein is intended to provide a consistent interpretation of 49 CFR §193.2067(b)(2) so that the wind design for "all other LNG facilities" meets the 49 CFR Part 193 requirements.

Commentary

As delineated in Part 193, Subpart C, each LNG facility designed after March 31, 2000, must comply with the requirements of Subpart C, as well as NFPA 59A (2001) (incorporated by reference in 49 CFR 193.2013) for LNG facility wind design. NFPA 59A, Section 4.1.4 specifies "the wind and snow loads for the design of LNG storage containers shall be determined using the procedures outlined in ASCE 7, *Minimum Design Loads and Associated Criteria for Buildings and Other Structures*. Where a probabilistic approach is used, a 100-year mean occurrence interval shall be used." The structural standard referenced in NFPA 59A (2001) for determining the wind load for the design of LNG storage containers is ASCE 7-93 (ASCE 1993). However, §193.2101 clarifies that if there is a conflict between Part 193 and NFPA 59A (NFPA 2001), the requirements in Part 193 prevail. Therefore, the wind forces for LNG facilities, including LNG storage containers, must be based on the requirements in §193.2067. At the time of the publication of this guidance, ASCE 7-05 was the referenced standard only for shop-fabricated containers of LNG or other hazardous fluids less than 70,000 gal. as prescribed in 49 CFR §193.2067(b)(1). Furthermore, the wind forces for "all other LNG facilities" must be based on a wind velocity as specified in 49 CFR 193.2067(b)(2). All LNG facility operators must demonstrate that the wind design for "all other LNG facilities" meets the requirements of 49 CFR §193.2067(b)(2). The wind design procedure as described Section 4.6.3 is expected to meet the intended 49 CFR Part 193 requirements for "all other LNG facilities."

The requirements set forth in 49 CFR §193.2067 specify a 150 mi/h sustained wind speed for all LNG facilities with the exception of shop-fabricated containers of LNG or hazardous fluids with capacities of less than 70,000 gal. However, the regulations do not prescribe the design load factors, importance factor, or other design parameters, such as shape coefficients or shielding for the facility and its structures. Among these variables are most notably the load factor and importance factor, which are the main components of the design wind force development that provide a factor of safety between the code-prescribed minimum design loads and the point of limit state exceedance. Typical limit states include load capacity of structural sections based on materials of construction and maximum deflection under loading. In consideration of these limit states, the LNG facility must be designed to withstand the direct effect of wind forces without the loss of structural or functional integrity. The engineer of record, in consultation with the facility operator, shall establish these parameters in support of the overall design basis with respect to wind loading.

4.6.3 Recommended Wind Design for LNG Facilities

The term *LNG facilities* in this section means "all other LNG facilities" as required in 49 CFR §193.2067(b)(2).

4.6.3.1 49 CFR §193.2067(b)(2) Wind Speed Applicability

The term *LNG facility* is defined in 49 CFR §193.2007 as "a pipeline facility that is used for liquefying natural gas or synthetic gas or transferring, storing, or vaporizing liquefied natural gas."

49 CFR §193.2067(b)(2) applies to all LNG facilities other than those described in §193.2067(b)(1).

Per PHMSA's frequently asked questions (FAQs), LNG facilities include

> The parts of the LNG plant considered for compliance with this design requirement include all parts used when liquefying natural gas or transferring, storing, or vaporizing liquefied natural gas. This includes piping and any equipment, facility, or building used in the transportation of gas or in the treatment of gas during the course of transportation.

Prior to implementation, the LNG facility operator should consult with PHMSA for clarification on the structures, buildings, equipment, and other factors to which the current paragraph of 49 CFR §193.2067(b)(2) applies. Careful consideration must be given to the impacts of non-LNG structures failing and the potential consequences to the LNG facility. Some operators have chosen to use a single wind speed for all facilities to mitigate these consequential damages. That being said, 49 CFR §193.2067(b)(2) is not applicable to structures, buildings, equipment, and piping that are not involved with liquefying natural gas or transferring, storing, or vaporizing liquefied natural gas, such as

- Gas pipeline entering or exiting facility and metering station,
- LNG facilities used in the course of natural gas treatment or hydrocarbon extraction that do not store LNG,
- Administration building,
- Maintenance and warehouse building,
- Guard houses, and
- Illumination towers.

4.6.3.2 Basic Design Wind Speed for LNG Facilities

To meet the regulatory requirements, either a sustained wind speed of 150 mi/h as required by 49 CFR §193.2067(b)(2)(i) or the wind speed corresponding to 0.5%

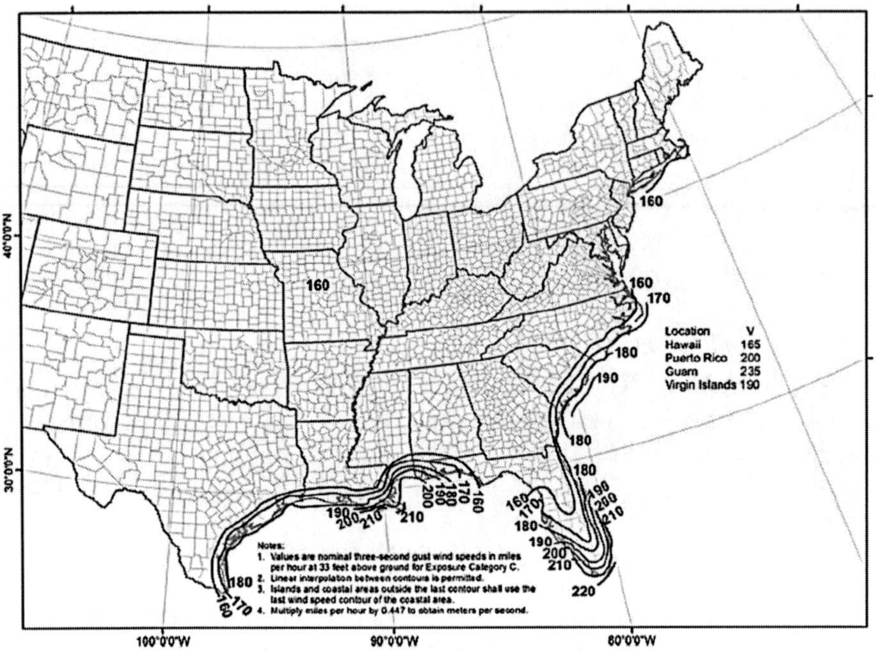

Figure 4-20. 10,000-year mean recurrence interval wind speed.
Source: ICC/NSSA *Standard for the Design and Construction of Storm Shelters*, ICC 500.

probability of exceedance in 50 years [mean recurrence interval (MRI) equal to 10,000 years] as required by 49 CFR §193.2067(b)(2)(ii) can be utilized in the design of LNG facilities. The specified 150 mi/h sustained wind speed is a strength-level wind speed (reflecting a load factor of 1.0) and should be properly converted to correspond with the applicable code. The following sections discuss wind speed conversions for different editions of ASCE 7.

A 3 s gust basic wind speed map with 10,000-year MRI was produced for ICC 500 (2014) by Applied Research Associates, using the same methodology as the ASCE 7 wind speed map. The 10,000-year MRI map is shown Figure 4-20. The basic wind speeds are peak 3 s gusts at 33 ft over flat, open (Exposure C) terrain. The 10,000-year MRI wind speed is also a strength-level wind speed, for which a wind load factor of 1.0 is applicable.

A sustained wind speed of 150 mi/h can be used to satisfy the requirements of 49 CFR §193.2067(b)(2). *Sustained* is defined as a 1 min average according to ASCE 7. For design purposes, a sustained wind speed of 150 mi/h needs to be converted to a 3 s gust wind speed. Depending on the LNG facility location, the conversion is performed using either the Durst curve or the Engineering Science Data Unit (ESDU) curve from Figure C26.5-1 of ASCE 7, which is repeated in Figure 4-21. Per C26.5.3 of ASCE 7, the ESDU curve has been shown to be valid

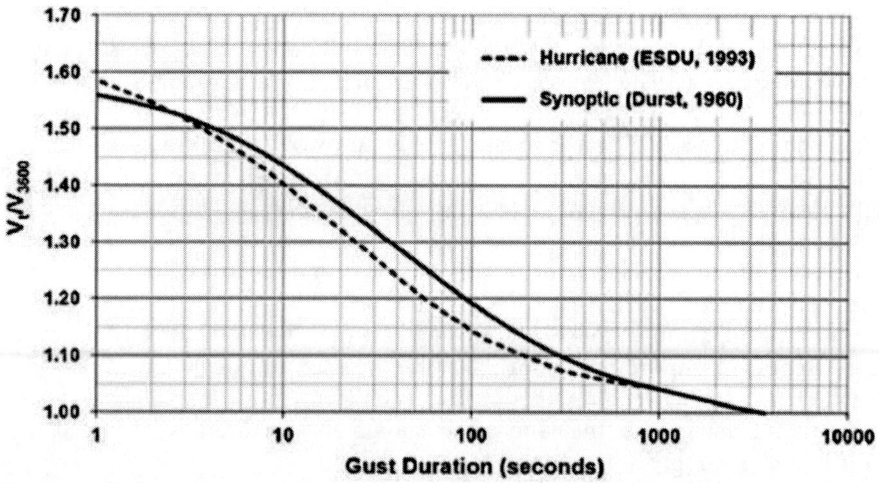

Figure 4-21. Maximum speed averaged over t_s *to hourly mean speed.*
Source: ASCE 7, Figure C26.5-1.

for hurricane winds. For hurricane winds, a sustained wind speed of 150 mi/h converts to a 3 s gust of 192 mi/h using the ESDU curve. For nonhurricane winds, a sustained wind speed of 150 mi/h converts to a 3 s gust of 183 mi/h using the Durst curve. Most LNG facilities in the United States are located in hurricane-prone regions. ASCE 7-05 does not reference the ESDU curve and only provides the Durst curve. PHMSA FAQ D1 states, "The Durst curve is an acceptable conversion method in ASCE/SEI 7-05, Chapter C6. Using this method, a sustained wind velocity of 150 mph is equivalent to a 183 mph 3-second gust."

4.6.3.3 Recommended ASCE 7-16 Wind Design Method for LNG Facilities

To meet the wind design requirements of 49 CFR §193.2067 for LNG facilities, the recommended design procedure using ASCE 7-16 is the strength design method. Either of the approaches described in Section 4.6.3.2 to establish the basic wind speed can be used to determine the design wind load for use with the strength design load combinations as shown in Section 2.3.1 of ASCE 7-16.

The allowable stress design method using the load combinations as shown in Section 2.4.1 of ASCE 7-16 can also be used to satisfy the requirements of 49 CFR §193.2067. However, to have an allowable stress design in ASCE 7-16 equivalent to the design using the strength design method in ASCE 7-05, the wind load factor as shown in Section 2.4.1 of ASCE 7-16 needs to be changed from 0.6 to 0.625 (1/1.6 = 0.625).

4.6.3.4 Recommended Procedures for Calculating Design Wind Forces

49 CFR §193.2067(b)(2) specifies that the wind forces for LNG facilities must be based on a sustained wind velocity of 150 mi/h or the wind velocity having a

probability of exceedance in a 50-year period of 0.5% or less. However, Part 193 does not prescribe a methodology to calculate the design wind forces. Because the wind speed as listed in 49 CFR §193.2067(b)(2) is a strength-level wind speed and ASCE 7-05 uses a service-level wind speed together with an importance factor and a load factor to compute wind forces, the wind speed for use with ASCE 7-05 must be adjusted accordingly. This can be accomplished by either converting the strength-level wind speed to a service-level wind speed by dividing by the square root of the ASCE 7-05 load factor ($V/\sqrt{1.6}$ or by modifying the wind load factor in all load combinations to be similar to those recommended for ASCE 7-16. The former approach may be less prone to error or misinterpretation because ASCE 7-05 can then be used directly. In both approaches, an importance factor of 1.0 can be used, although some operators elect to use a higher factor. Other factors, such as the flexible gust effect factor, whose value depends on the wind speed, should also be carefully evaluated and adjusted accordingly.

4.6.3.5 Commentary on the Requirements of 49 CFR §193.2067(a)(3)

49 CFR §193.2067(a)(3) requires that impounding systems for LNG storage tanks be designed to withstand impact forces and potential penetrations by wind-borne missiles without loss of structural or functional integrity.

For compliance with Part 193, a rational method satisfying the requirements of 49 CFR §193.2067(a)(3) needs to be demonstrated and submitted to PHMSA by the LNG facility operator.

It is not uncommon to see earthen dikes for the impounding systems of LNG storage tanks. If the earthen dikes are sufficiently wide, it may not be necessary to perform impact and penetration analysis. In these cases, engineering judgment is required.

For the design of other impounding system dikes (like concrete) to withstand impact forces and potential penetrations by wind-borne missiles, some engineers have used finite-element analysis methods to demonstrate the adequacy of impounding system dikes. However, no specific requirements regarding missiles to be used for evaluating the adequacy of impoundment systems are provided in the 49 CFR §193.2067 requirements. Some other available references that can be useful for establishing wind-borne missile design and evaluation criteria are as follows:

- Information for some standard wind-borne missile types and velocities can be found in Table 4 of ANSI/ANS-2.3 (2011). Similar information is also available in NRC 1.76 (2007) and NRC 1.221 (2011).
- Reinforced concrete wall design owing to impact load can be found in Section 8.4 of ACI 376 (2011) and CEB Bulletin 187 (1988). FERC currently prefers the missiles impact procedures of CEB Bulletin 187 to those in Section 8.4 of ACI 376-11.
- Design of concrete structures for impulsive and impactive loading can be found in Appendix F of ACI 349 (2013).

4.7 EVALUATION OF WIND LOADS ON EXISTING STRUCTURES

4.7.1 Introduction

In a petrochemical facility, it is common for process structures and pipe racks to be altered or modified from their original installation. Proposed changes and modifications to an existing facility must be evaluated for their impact on existing equipment support structures and foundations.

The engineering evaluation of existing structures and foundations should generally be approached in a different manner than the design of new facilities. Conservative design loads and a conservative application of the design loads are common practices in the design of new facilities. However, this same approach may not be warranted when evaluating existing structures. Conservatism in new design generally has a minimal impact on the installed cost of the structure, but the same conservatism in the evaluation of an existing structure can have a significant negative cost and schedule impact on a project and may result in unnecessary upgrades.

The following sections provide references and summaries of current industry guidelines for the evaluation of existing structures and foundations.

4.7.2 References to Industry Codes and Guidelines

Several model building codes and industry guidelines address additions and alterations to existing buildings and structures. The following references provide some guidance regarding the evaluation and application of wind loading on modified structures.

The following references use either a 5% guideline (generally for gravity loads) or a 10% guideline (for lateral loads) for determining when upgrades of structural elements are required. When the modification does not increase the force (or increase the demand-to-capacity ratio) in an existing structural element by more than 5% or 10%, whichever is applicable, then a structural upgrade of that element is not required.

The engineer is referred to Section 4.7.4.2 for a caution regarding the need to consider any past modifications (or structural upgrades) when evaluating the impact of a new modification.

4.7.2.1 PIP STC01015 Structural Design Criteria, April 2017

Section 4.4 states:

If the owner and engineer of record agree that the integrity of the existing structure is 100% of the original design and the existing loading on the structure is 100% or less of the original allowable capacity based on the materials used and design codes in effect at the time of the original design, structural designs should be performed in accordance with the following:

(a) If additions or alterations to an existing structure do not increase the force in any structural element or connection by more than 5%, no further analysis is required.

(b) If the increased forces on the element or connection are greater than 5%, the element or connection shall be analyzed to show that it is in accordance with the applicable design code for new construction.

(c) The strength of any structural element or connection shall not be decreased to less than that required by the applicable design code or standard for new construction for the structure in question.

4.7.2.2 International Building Code 2018

Chapter 34, "Existing Structures," was removed from the IBC in the 2012 revision. Previous versions contained definitions for modifications and additions as noted in the following, but in the latest version this as all been moved to the International Existing Building Code (IEBC).

In IBC 2006, (ICC 2006) modifications (alteration, repair, or addition) to existing buildings are addressed in Chapter 34, "Existing Structures." Section 3403.2 states "Additions or alterations to an existing structure shall not increase the force in any structural element by more than 5 percent, unless the increased forces on the element are still in compliance with the code for new structures, nor shall the strength of any structural element be decreased to less than that required by this code for new structures."

In IBC 2009 (ICC 2009), additions to existing buildings are addressed in Chapter 34 Section 3403, and existing structural elements are addressed in Sections 3403.3 and 3403.4. The provisions in Section 3403.3 for members carrying gravity load are the same as for IBC 2006. However, the provisions in Section 3403.4 are new and address members carrying lateral load. Section 3403.4 states, "Any existing lateral load-carrying structural element whose demand-capacity ratio with the addition considered is no more than 10 percent greater than its demand-capacity ratio with the addition ignored shall be permitted to remain unaltered."

4.7.2.3 International Existing Building Code 2018

IEBC (ICC 2018b) uses the same guideline of 5% for vertical members and 10% for lateral load-carrying members recommended in IBC 2009 Section 502.4, which states, "Any existing gravity load-carrying structural element for which an addition and its related alterations cause an increase in design dead, live, or snow load, including snow drift effects, of more than 5% shall be replaced or altered as needed to carry the gravity loads required by the International Building Code for new structures. Any existing gravity load-carrying structural element whose vertical load-carrying capacity is decreased as part of the addition and its related alterations shall be considered to be an altered element subject to the requirements of Section 503.3. Any existing element that will form part of the lateral load path for any part of the addition shall be considered to be an existing lateral load-carrying structural element subject to the requirements of Section 502.5." In addition, Section 502.5, Exception 1 states, "Any existing lateral load-carrying structural element whose demand-capacity ratio with the addition considered is not more than 10 percent greater than its demand-capacity ratio with the addition ignored shall be permitted to remain unaltered."

For purposes of work classification, changes to existing buildings are defined in IEBC (2018b) as follows:

- Repairs: "The reconstruction replacement or renewal of any part of an existing building for the purpose of its maintenance or to correct damage." Repairs include patching, restoration, or replacement of damaged materials. Buildings that have sustained substantial structural damage to the vertical elements of the lateral-force-resisting system shall be evaluated. The wind design level for repair shall be as required by the building code in effect at the time of the original construction, unless the damage was caused by wind, in which case the design level shall be as required by the code in effect at the time of the original construction or as required by IBC, whichever is greater.
- Alterations: "Any construction or renovation to an existing structure other than a repair or addition." Alterations do not generally affect wind forces unless there are additions or eliminations of doors and windows that change the closed or open classification of the building and thus affect the wind forces on the building. Alterations shall be such that the load-carrying capacity of the existing structure is not reduced.
- Additions: "An extension or increase in floor area, number of stories, or height of a building structure."
  - Additions to existing buildings are new construction and are required to comply with the IBC. Any elements of the existing lateral-force-resisting system subjected to increased lateral or vertical loads due to vertical or horizontal additions are required to comply with the IBC.

The requirements described are the same in IEBC (2015), but the chapters and sections were revised in the 2018 version.

4.7.3 Considerations for Wind Exposure Category

The design of new petrochemical structures is generally based on Wind Exposure Category C [open terrain with scattered obstructions that have heights generally less than 30 ft (9.1 m)] or Category D (flat, unobstructed areas and water surfaces). However, many existing pipe racks and equipment structures are located in areas of the plant that are more representative of Exposure Category B.

For the evaluation of existing structures and foundations, a less conservative wind exposure category may be considered. The exposure category for a particular location should be established in accordance with ASCE 7, and it should be noted that the standard permits interpolation between exposure categories (Section 26.7.3, "Exposure Categories"). Examples that provide guidance on simplified interpolation methods are provided in C26.10.1.

For each wind direction considered, an exposure category that adequately reflects the characteristics of ground roughness and surface irregularities should be determined for the site at which the structure is located. For some structures, it would be appropriate (but not required) to use a different exposure category for different wind directions.

Caution: Care should be exercised when establishing the exposure category for structures near the plant boundary. Generally, the area outside the fence line should be considered as characteristic of Exposure Categories C or D.

The following items highlight additional considerations when determining the wind exposure category to be used when evaluating existing structures. ASCE 7-16 C26.7 provides additional detail on calculation of an exposure category.

Surface Roughness B is defined as urban and suburban terrain with numerous, closely spaced obstructions having the size of single-family dwellings or larger. To qualify as Exposure Category B, a ground surface roughness, as defined by Surface Roughness B, must prevail for a distance of 2,600 ft or 20 times the maximum height of the structure in the upwind direction, whichever is greater. There is also an exception when the mean height is less than or equal to 30 ft, and in this case the upwind distance may be reduced to 1,500 ft (ASCE 7, Section 26.7.3).

Most petrochemical facilities comprise numerous, closely spaced vessels and structures, but usually have many structures in excess of 30 ft high. The use of Exposure Category B should be considered when the location relative to upwind surface roughness would allow. Figure 4-22 shows a petrochemical facility and a typical plant layout with closely spaced vessels, pipe racks, storage tanks, and equipment structures.

For comparison purposes, the wind pressure for Exposure Category B is approximately 70% of the pressure for Exposure Category C at 30 ft elevation.

4.7.4 Other Cautions and Considerations

4.7.4.1 Review of Design Basis

Some plant sites are under the jurisdiction of local building officials. For these locations, the basis for the engineering evaluation should be reviewed and meet the requirements of the local building officials. For all sites, the basis for the engineering evaluation should always be reviewed up front with the owner of the facility.

4.7.4.2 Potential for Past Modifications

Existing structures may have been modified since their original installation. Over the years, additional equipment may have been added to a structure, or the original structure may have been expanded horizontally or vertically. An overall gravity, wind, or seismic evaluation of the structure may not have been performed at the time of a past modification. A review of the existing drawings can provide valuable

Figure 4-22. Typical petrochemical facility.
Source: Courtesy of FEMA.

information regarding past modifications and also provide information regarding any structural upgrades that were made at the time of the additions or modifications.

It is also possible that the original structure has been modified and potentially weakened, either by the removal of horizontal or vertical bracing members or by the modification of individual beams or columns. Figure 4-23 captures an unusual field modification to an existing steel column in a structure.

The available engineering drawings may not accurately represent actual field conditions. A site visit to the plant should be conducted to verify the as-built condition of all critical members in the existing structure, especially those that are directly impacted by the proposed modification.

4.7.4.3 Potential for Structural Deterioration

Existing structures in the plants are subject to harsh environments and may have experienced some deterioration and loss of strength. A site visit should be made to assess the current condition of an existing steel or concrete structure. During the site visit, the engineer should look for signs of active corrosion, especially for signs of corrosion that may be somewhat hidden under fireproofing. Figure 4-24 shows typical cracking and deterioration of a concrete column, which indicates underlying corrosion of the reinforcing steel.

A visual-only assessment may not be adequate to assess the condition of the existing structure, and it may be necessary to conduct a formal condition survey of any questionable members. There are firms that specialize in conducting condition surveys, and their services should be utilized as appropriate.

Figure 4-23. Field modification to an existing steel column.

Figure 4-24. Deterioration of a concrete column.

4.8 WIND LOAD ANALYSIS UNCERTAINTY

Many of the structures that appear in petrochemical and other industrial facilities have unique and complex geometries that have received relatively little attention from wind engineering researchers. As such, it would seem that a designer's confidence in the accuracy of the available methods for estimating wind loads on these types of structures would be somewhat lower than for structures typically appearing in commercial or residential settings.

Amoroso and Levitan (2009a, b) compiled available wind tunnel test results for model open-frame structures and vertical vessels to assess the uncertainty of the wind load analysis methods available for these structure types. The values of the predicted force coefficients were compared to the measured results for each of 48 different model configurations. The predicted coefficients were calculated using the provisions from Chapter 4 of the first edition of this guide (ASCE 1997) and also using the provisions contained in Chapter 5 of ASCE (2011).

The predictions using both the original and updated provisions were found to slightly overestimate the wind loads for the model structures. The standard deviations of the ratios of the measured values to the predicted values were found to be comparable to the variability associated with net pressure coefficient estimation for typical enclosed building structures (Ellingwood and Tekie 1999). The standard deviation for the estimates using the updated guide provisions were nearly one-third lower than for the estimates using the original guide provisions, indicating a reduction in overall uncertainty upon application of the updated provisions.

Although the results of this analysis ultimately showed that the probability of failure for a petrochemical structure under the action of wind may be slightly higher than for a regular building structure, the differences were indeed small. Furthermore, the main reason for the difference in overall reliability between the modeled petrochemical structures and regular buildings was that the net pressure coefficient estimates for regular buildings are systematically overestimated to a greater degree.

The findings reported in this section were based on wind tunnel models of comparatively simple open-frame structures and vessels. The data set did not include any very large and/or highly complex structures.

References

ACI (American Concrete Institute). 2011. *Code requirements for design and construction of concrete structures for the containment of refrigerated liquefied gases and commentary.* ACI 376. Farmington Hills, MI: ACI.

ACI. 2013. *Code requirements for nuclear safety-related concrete structures and commentary.* ACI 349. Farmington Hills, MI: ACI.

Amoroso, S., and M. Levitan. 2009a. "Recent research for wind loads on petrochemical structures." In *Proc., ASCE/SEI 2009 Structures Congress,* Austin, TX, 2088–2097.

Amoroso, S., and M. Levitan. 2009b. "Wind load analysis uncertainty for petrochemical structures." In *Proc., 11th Americas Conf. on Wind Engineering,* San Juan, Puerto Rico.

ANSI (American National Standards Institute). 2011. *Estimating tornado, hurricane, and extreme straight line wind characteristics at nuclear facility sites.* ANSI/ANS-2.3. Washington, DC: ANSI.

ASCE. 1993. *Minimum design loads for buildings and other structures.* ASCE 7-93. Reston, VA: ASCE.

ASCE. 1997. *Wind loads and anchor bolt design for petrochemical facilities.* Reston, VA: ASCE.

ASCE. 2005. *Minimum design loads for buildings and other structures.* ASCE 7-05. Reston, VA: ASCE.

ASCE. 2011. *Wind loads for petrochemical and other industrial facilities.* Reston, VA: ASCE.

ASCE. 2012. *Wind tunnel testing for buildings and other structures.* ASCE 49-12. Reston, VA: ASCE.

ASCE. 2014. *Design loads on structures during construction.* ASCE 37-14. Reston, VA: ASCE.

ASCE. 2017. *Minimum design loads and associated criteria for buildings and other structures.* ASCE 7-16. Reston, VA: ASCE.

Calvert, E., and F. H. Fouad. 2000. "A review of current wind load provisions for transmission pole structures." In *Proc., Structures Congress 2000,* Philadelphia.

CEB (Comité Euro-International du Béton). 1988. *Concrete structures under impact and impulsive loading.* Bulletin 187. Lausanne, Switzerland: CEB.

Ellingwood, B. R., and P. B. Tekie. 1999. "Wind load statistics for probability-based structural design." *J. Struct. Eng.* **125** (4): 453–463. https://doi.org/10.1061/(ASCE)0733-9445(1999)125:4(453).

Godoy, L. A. 2007. "Performance of storage tanks in oil facilities damaged by Hurricanes Katrina and Rita." *J. Perform. Constr. Facil.* **21** (6): 441–449. https://doi.org/10.1061/(ASCE)0887-3828(2007)21:6(441).

ICC (International Code Council). 2006. *International building code (IBC).* Washington, DC: ICC.

ICC. 2009. *International building code (IBC).* Washington, DC: ICC.

ICC. 2014. *Standard for the design and construction of storm shelters.* ICC 500/NSSA or ICC 500. Washington, DC: ICC.

ICC. 2015. *International existing building code (IEBC).* Washington, DC: ICC.

ICC. 2018a. *International building code (IBC).* Washington, DC: ICC.

ICC. 2018b. *International existing building code (IEBC).* Washington, DC: ICC.

IEEE. 1996. *National electrical safety code.* NESC C2-1997. New York: IEEE.

J&H Marsh & McLennan. 1998. *Large property damage losses in the hydrocarbon-chemical industries—A thirty year review, risk control strategies.* 18th ed. Fort Lauderdale, FL: J&H Marsh & McLennan.

NFPA (National Fire Protection Association). 2001. *Standard for the production, storage, and handling of liquefied natural gas (LNG).* NFPA 59A. Quincy, MA: NFPA.

NRC (Nuclear Regulatory Commission). 2007. *Design-basis tornado and tornado missiles for nuclear power plants.* NRC 1.76. Washington, DC: NRC.

NRC. 2011. *Design-basis hurricane and hurricane missiles for nuclear power plants.* NRC 1.221. Washington, DC: NRC.

PIP (Process Industry Practices). 2017. *Structural design criteria.* PIP STC01015. Austin, TX: PIP.

CHAPTER 5

Recommended Guidelines Part II: Analytical Determination of Wind Loads

This committee has updated the recommended guidelines and commentary in this chapter to reflect recent updates to ASCE 7.

GUIDELINES

5.1 General

Design wind forces for the main wind-force-resisting system (MWFRS) and components should be determined by the equation

$$F = q_z\ G\ C_f\ A \qquad (5\text{-}1a)$$

(where F is the applied wind force) using the following procedure:

1. The velocity pressure, q_z, is determined in accordance with the provisions of ASCE 7, Section 26.10.
2. The gust effect factor, G (or G_f), is determined in accordance with the provisions of ASCE 7, Section 26.11.

COMMENTARY

C5.1 General

The basic equation for design wind loading [Equation (5-1a)] is adopted from ASCE 7 procedures for "Design Wind Loads on Other Structures" [Section 29.4, Equation (29.4-1)]. The provisions in this chapter primarily provide guidance in selecting appropriate force coefficients and projected areas.

Variable G_f is used in place of G for flexible structures, defined by ASCE 7 as structures with a fundamental frequency $f < 1$ Hz. If the height divided by the least horizontal dimension is greater than 4, a frequency check may be warranted.

Procedures and guidance for computing gust effect factors are provided in Section 26.11 and commentary of the standard.

3. Force coefficients C_f and corresponding projected areas A_f or A_e are determined from the provisions of Section 5 for structures and equipment.

5.2 Pipe Racks

Wind on the pipe rack should be calculated as outlined in the following. The provisions apply to transverse wind loads on the pipe rack.

C5.2 Pipe Racks

It is the experience of this committee that pipe friction and anchor loads govern the longitudinal design loads of a typical pipe rack. However, the engineer should exercise judgment to ensure longitudinal wind loads are considered for unusual pipe rack configurations.

5.2.1 Tributary Area for Piping

The tributary area for piping should be based on the diameter of the largest pipe, *D*, plus 10% of the width of the pipe rack, *W*. This result is multiplied by the length of the pipes (bent spacing, *L*) to determine the tributary area,

where

D = Largest pipe diameter in bent (including insulation),

W = Width of the bent,

L = Bent spacing,

A = Tributary area, and

$A = L(D + 10\%W)$.

C5.2.1 Tributary Area for Piping

This area is based on the assumption that the wind will strike at an angle of +5.7° or − 5.7° from the horizontal (e.g., a plus or minus slope of 1 to 10) and that the largest pipe is on the windward side. In some cases, the pipe rack longitudinal strut or stringer might fall in the shielding envelope and could be deleted from the wind load considerations.

This is a reasoned approach that accounts for wind on all the pipes (or cable trays) and shielding of the leeward pipes (or cable trays). The basis for the selection was a review of existing practices prior to publication of the previous edition (ASCE 2011) of this book.

5.2.2 Tributary Area for Cable Trays

The tributary area for cable trays should be based on the height of the largest tray, h, plus 10% of the width of the pipe rack, W. This result is multiplied by the length of the trays (bent spacing, L) to determine the tributary area,

where
h = Height of cable tray,
W = Width of bent,
L = Bent spacing,
A_c = Area of cable tray, and
$A_c = L(h + 10\%W)$.

C5.2.2 Tributary Area for Cable Trays

See C5.2.1.

5.2.3 Force Coefficients for Structural Members

For all structural members, $C_f = 1.8$, or alternatively $C_f = 2.0$ at and below the first level and $C_f = 1.6$ for members above the first level. No shielding shall be considered for the leeward column line except as noted in C5.2.1.

C5.2.3 Force Coefficients for Structural Members

The C_f value was determined with guidance from ASCE 7-16 Figure 29.4-2 considering typical solidities above and below the first level. For pipe rack structures with more than two column lines, shielding may be considered.

5.2.4 Force Coefficient for Pipes

The force coefficient $C_f = 0.7$ should be used as a minimum.

C5.2.4 Force Coefficient for Pipes

The force coefficient C_f for pipe is taken from ASCE 7-16 Figure 29.4-1 for a round shape, with $h/D = 25$, $D\sqrt{q_z} > 2.5$, and a moderately smooth surface; that is, $C_f = 0.7$. If the largest pipe is insulated, then consider using C_f for a rough pipe dependent on the roughness coefficient of the insulation (D'/D).

5.2.5 Force Coefficient for Cable Trays

For cable trays, the force coefficient $C_f = 2.0$.

C5.2.5 Force Coefficient for Cable Trays

The force coefficient C_f for cable trays is taken from ASCE 7-16 Figure 29.4-1 for a square shape with the face normal to the wind and with $h/D = 25$; that is, $C_f = 2.0$.

5.3 Open-Frame Structures

5.3.1 General

This section covers wind loads on open-frame or lattice structures, with or without equipment, piping, electrical items, stairs, ladders, and handrails, among others.

Wind loads should be calculated in accordance with the general procedures and provisions of ASCE 7-16 for wind loads on "Other Structures" with the exceptions as noted.

5.3.1.1 Main Wind-Force-Resisting System

Wind forces acting on the structural frame and appurtenances (e.g., ladders, handrails, stairs) should be computed in accordance with 5.3.2.

C5.3.1.1 Main Wind-Force-Resisting System

The methods used to calculate wind loads on an open-frame structure in 5.3 and Appendix 5A were adapted from theoretical work (Cook 1990), wind

A somewhat more exact and detailed method of calculating the wind forces on the structural frame and appurtenances is presented in Appendix 5A. This method allows the designer to estimate frame forces for a wider range of frame spacing and solidity ratios. This method also allows the designer to calculate the wind angle producing the maximum force.
A simplified method that addresses high-solidity structures (greater than 50% projected solidity including framing, equipment, and appurtenances) is presented in Appendix 5B.

tunnel testing results (Georgiou 1979), and related analytical work (Willford and Allsop 1990, Nadeem and Levitan 1997) dealing with wind forces on identical, regularly spaced frames made of sharp-edged members.

These methods have been extended to handle cases such as frames of unequal solidity, the presence of secondary beams (beams not along column lines), and frames made up of rounded members (Willford and Allsop 1990, Georgiou et al. 1981). Only limited experimental work has been carried out on some of these cases (Qiang 1998, Levitan et al. 2004, Amoroso and Levitan 2009). It is the opinion of the committee that it is not unreasonable to presume that for a structure that is not particularly unusual or irregular, or does not have too many appurtenances, the procedures of 5.3 and Appendix 5A should yield wind load results for the structure and the appurtenances together with a reliability that is consistent with the other provisions in this chapter.

Research by Amoroso and Levitan (2011) shows that the force coefficients for open-frame structures with high projected solidity ratios can be reasonably estimated by considering a small number of key variables related to the overall geometry of the open-frame structure. This approach does not require the designer to consider the combination of effects from framing, equipment, and other elements. This method is presented in Appendix 5B.

Wind forces on equipment, piping and cable trays located on or attached to the structure should be calculated according to the applicable provisions of Chapter 5 and added to the wind forces acting on the frame in accordance with 5.3.6.

Scarce experimental work to date (Qiang 1998, Levitan et al. 2004, Amoroso and Levitan 2009) has considered the inclusion of three-dimensional solidity (e.g., vessels, heat exchangers) placed in the framework. In general, it is expected that the total wind load on equipment will be less than the sum of the loads on the individual items owing to shielding by the frame and other neighboring equipment.

Thus, the approach taken in 5.3.6 is the reduction of the total wind load on equipment by a multiplication factor η_{equip} to account for this shielding.

5.3.1.2 Force Coefficients for Components

Wind loads for the design of individual components, cladding, and appurtenances (excluding equipment, piping, and cable trays) should be calculated according to the provisions of ASCE 7-16. Based on common practice, force coefficients and areas for several items are given in Table 5-1. Force coefficients for open-frame structures with partial cladding are given in 5.4.

5.3.2 Frame Load

For open-frame structures, design wind forces for the main wind-force-resisting system should be determined by the equation

$$F_S = q_z\ G\ C_f\ A_e \qquad (5\text{-}1b)$$

C5.3.2 Frame Load

The structure is idealized as two sets of orthogonal frames. The maximum wind force on each set of frames is calculated independently.

Note: In this equation, C_f accounts for the entire structure in the direction of the wind.

Table 5-1. Force Coefficients for Wind Loads on Components.

Item	C_f	*Projected area*
Handrail including toe board	2.0	0.80 ft²/ft
Ladder without cage	2.0	0.50 ft²/ft
Ladder with cage	2.0	0.75 ft²/ft
Solid rectangle and flat plate	2.0	—
Round or square shape	See ASCE 7-16, Figure 29.4-1	—
Stair with handrail		
Side elevation	2.0	Handrail area plus channel depth
End elevation	2.0	50% of gross area

In Equation (5-1b), F_S is the wind force on the structural frame and appurtenances, q_z and G are as defined in Section 5.1, and

- The force coefficient C_f is determined from the provisions of 5.3.3.
- The area of application of force A_e is determined per 5.3.5.
- The design load cases are computed per 5.3.6.

5.3.2.1 Limitations of Analytical Procedure

Design wind forces are calculated for the structure as a whole.

The method is described for structures that are rectangular in plan and elevation.

C5.3.2.1 Limitations of Analytical Procedure

No information is provided about distribution of loads to individual frames. However, the windward frame will experience a larger percentage of the total wind force than any other frames, except possibly for the case where the solidity ratio of the windward frame is much less than that of other frames.

5.3.3 Force Coefficients

The force coefficient for a set of frames shall be calculated by

$$C_f = C_{Dg}/\varepsilon \qquad (5\text{-}2)$$

where C_{Dg} is the force coefficient for the set of frames given in Figure 5-1, and ε is the solidity ratio calculated in accordance with 5.3.4.

Extrapolation of the charts is required for frame solidity ratios greater than 0.35.

Alternately, C_f may be determined using Appendix 5A.

C5.3.3 Force Coefficients

Force coefficients C_{Dg} are obtained from Figure 5-1 (see C5.3.1.1 for research references) or Appendix 5A. A single value is obtained for each axis of the structure. This value is the maximum force coefficient for the component of force acting normal to the frames for all horizontal wind angles. Although the wind direction is nominally considered as being normal to the set of frames under consideration, the maximum force coefficient occurs when the wind is not normal to the frames (see C5.3.6.1 and 5A.1). The angle at which the maximum force coefficient occurs varies with the dimensions of the structure, solidity, number of frames, and frame spacing.

Force coefficients are defined for wind forces acting normal to the frames irrespective of the actual wind direction.

A method to estimate this angle is given in Appendix 5A, which also provides C_f values for a larger range of S_F/B and ε values than Figure 5-1.

The frame spacing ratio is equal to S_F/B, where S_f is the frame spacing in the direction parallel to the wind and B is the frame width as shown in Figure 5-1.

Extrapolation of the charts is required for frame spacing ratios greater than 0.5.

The force coefficients C_{Dg} were developed for use on the gross area (i.e., envelope area) of the structure as used by the British wind loading standard (Willford and Allsop 1990). These are converted to force coefficients that are applied to solid areas as used in ASCE 7-16 by Equation (5-2).

The force coefficients C_{Dg} were developed from wind tunnel tests for structures with a vertical aspect ratio (ratio of height to width perpendicular to the flow direction) of 4.

Although vertical aspect ratio does not play a large role in determining overall loads, the coefficients given in Figure 5-1 may be slightly conservative for relatively shorter structures and slightly unconservative for relatively taller structures.

Force coefficients C_{Dg} are applicable for frames consisting of typical sharp-edged steel shapes such as wide flange shapes, channels, and angles. Georgiou et al. (1981)

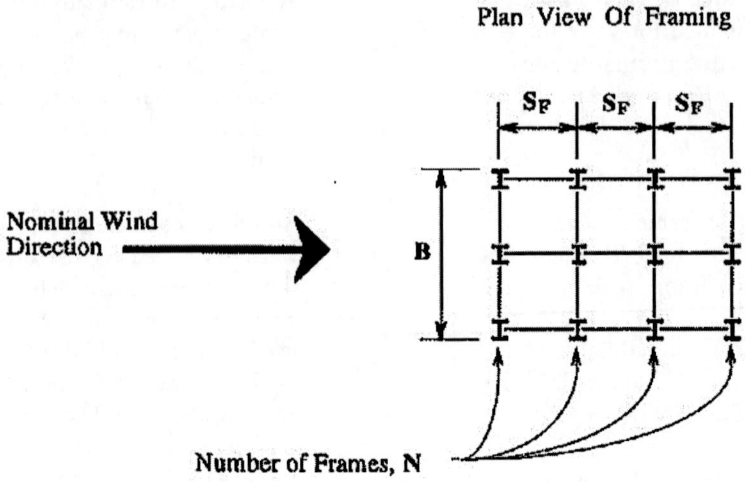

Notes:

1. Frame spacing ratio is defined as S_F/B.
2. Frame spacing, S_F, is measured from centerline to centerline.
3. Frame width, B, is measured from outside edge to outside edge.
4. Number of frames, N, is the number of framing lines normal to the nominal wind direction (N=4 as shown).
5. Linear interpolation may be used for values of S_F/B not given on the following pages.

Figure 5-1. Force coefficients, C_{Dg}, for open-frame structures.

suggest a method to account for structures containing some members of circular or other cross-sectional shape.

5.3.4 Solidity Ratio

The solidity ratio ε is given by

$$\varepsilon = A_S/A_g \qquad (5\text{-}3)$$

where A_g is the gross area (envelope area) of the windward frame, and A_S is the effective solid area of the windward frame

C5.3.4 Solidity Ratio

Willford and Allsop (1990) present a method to account for the effects of secondary floor beams (beams not in the plane of a frame). Use of the Willford and Allsop method may result in a small increase in

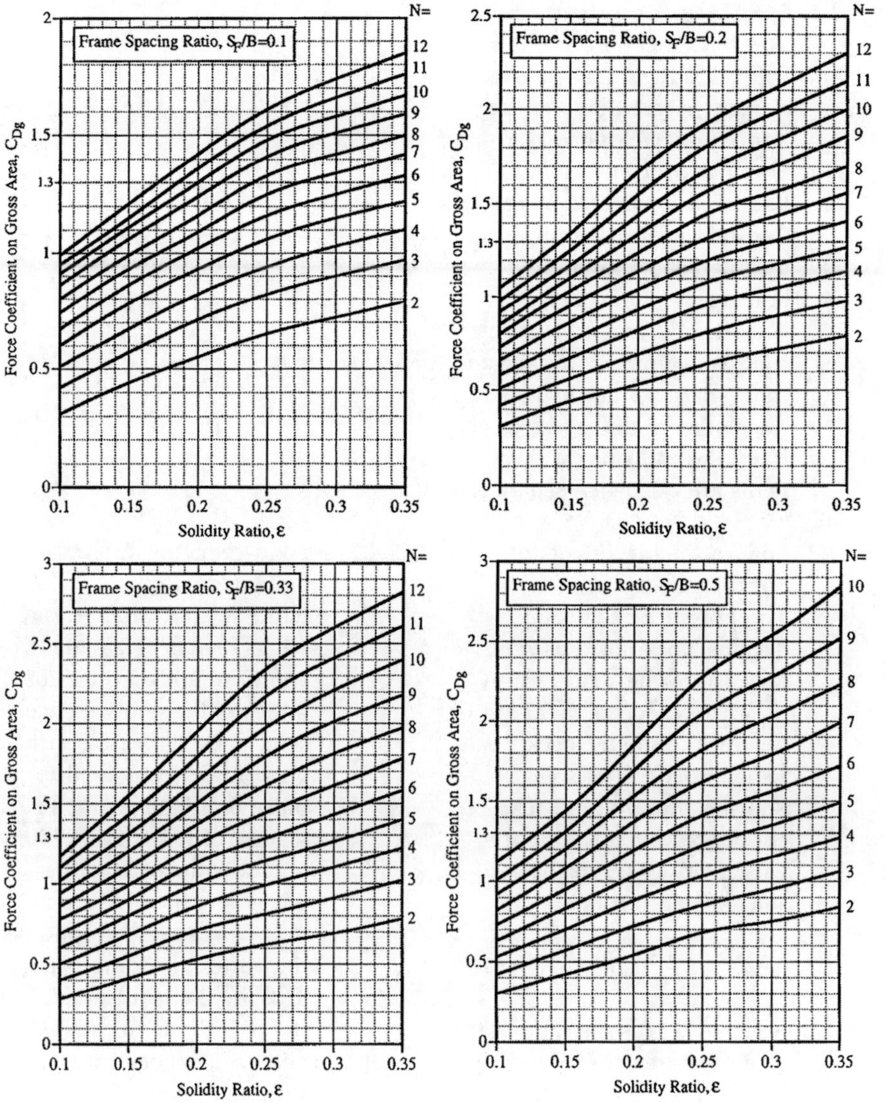

Figure 5-1. (Continued). Force coefficients, C_{Dg}, for open-frame structures.

defined by the following subsections:

the total wind force on the structure. With the associated uncertainties with the determination of the wind forces, this minor addition may be ignored.

5.3.4.1 Solid Area of a Frame

The solid area of a frame is defined as the solid area of each element in the plane of the frame projected normal to the nominal wind direction. Elements considered as part of the solid area of a frame include beams, columns, bracing, cladding, stairs, ladders, and handrails. Items such as equipment, piping, and cable trays are not included in calculation of the solid area of a frame; wind loads on these items are calculated separately.

5.3.4.2 Load-reducing Effects of Solid Flooring

The presence of flooring or decking does not cause an increase of the solid area of 5.3.4.1 beyond the inclusion of the thickness of the deck. Load-reducing effects of solid flooring (not grating) may be considered by adjusting the wind load by up to the shielding factor η_{floor}

$$\eta_{floor} = 1 - 0.2(A_{fb}/A_S) \quad (5\text{-}4)$$

where A_{fb}/A_S is the ratio of the projected area contributed by horizontal beams (in the vertical windward plane) supporting solid floors to the total projected solid area. This factor should be applied to the frame force coefficient.

C5.3.4.2 Load-reducing Effects of Solid Flooring

Experimental work (Qiang 1998, Levitan et al. 2004, Amoroso and Levitan 2009) indicates that the presence of solid decking decreases the wind forces compared to those of a bare frame. A history of research on bluff bodies with wake splitter plates reveals a reduction in drag owing to a disruption of the vortex shedding pattern. This mechanism may explain the reductions in drag observed for open frames with solid floors. The relationship for the reduction factor given here is an empirical representation of the reduction of the wind load as a function of the solid area provided by floor beams (Amoroso and Levitan 2009).

No research related to open-grating floors has been published. There is no evidence that open-grating floors will significantly affect the wind forces on the structure.

5.3.4.3 A_S for Structures with Frames of Equal Solidity

For structures with frames of equal solidity, the effective solid area A_S should be taken as the solid area of the windward frame.

5.3.4.4 A_S for Structures with Frames of Unequal Solidity – Case 1

For structures where the solid area of the windward frame exceeds the solid area of the other frames, the effective solid area A_S should be taken as the solid area of the windward frame.

C5.3.4.4 A_S for Structures with Frames of Unequal Solidity – Case 1

The force coefficients of Figure 5-1 were developed for sets of identical frames. Research shows that the solidity of the windward frame is the most critical (Cook 1990, Whitbread 1980), leading to the recommendation. This provision is likely to yield slightly conservative loads because the greater the solidity of the windward frame with respect to the other frames, the greater the shielding of the other frames.

5.3.4.5 A_S for Structures with Frames of Unequal Solidity – Case 2

For structures where the solid area of the windward frame is less than the solid area of the other frames, the effective solid area A_S should be taken as the average of all the frames.

5.3.4.6 A_S for Structures with Vertical Bracing Members

When vertical bracing members in frames parallel to the nominal wind direction are present, the vertical projected area normal to the nominal wind direction for the vertical bracing members shall be added to A_S. Regardless of the configuration of the vertical bracing (e.g., diagonal bracing, chevron bracing, K-bracing, X-bracing) or arrangement (bracing located totally in one bay or distributed among several bays of a bent), the vertical projected area of only one brace member per story per braced bent parallel to the wind direction shall be considered.

C5.3.4.6 A_S for Structures with Vertical Bracing Members

Experimental work (Amoroso and Levitan 2009) indicates that neglecting the contributions of vertical bracing to the solid area for wind directions nominally parallel to the plane of the bracing can lead to unconservative estimates of the wind load on an open-frame structure.

5.3.5 Area of Application of Force

The value of A_e shall be calculated in the same manner as the effective solid area in 5.3.4, except that it is for the portion of the structure height consistent with the velocity pressure q_z.

5.3.6 Design Load Cases

The total wind force acting on the structure in a given direction, F_T, is equal to the sum of the wind loads acting on the structure and appurtenances (F_S) plus the wind load on the equipment and vessels (per Section 5.5) plus the wind load on piping. See Figure 5-2 for complete definitions of F_T and F_S.

C5.3.6 Design Load Cases

In some cases, this design load will exceed the load that would occur if the structure were fully clad. It is also possible that the wind load on just the frame itself (before equipment loads are added) will exceed the load on the fully clad structure. This happens most often for structures with at least four or five frames and relatively

If piping arrangements are not known, the engineer may assume the piping area to be 10% of the gross area of the face of the structure for each principal axis. A force coefficient of 0.7 should be used for this piping area.

The two load cases shown in Figure 5-2 should be considered.

5.3.6.1 F_T and F_S

F_T **and** F_S are applied to each of the two design load cases as follows: Frame load + equipment load + piping load (F_T) for one axis, acting simultaneously with 50% of the frame load (F_S) along the other axis, for each direction.

The combination of wind with other loads shall be computed in accordance with ASCE 7-16 Section 2.0.

higher solidities. This phenomenon is very clearly demonstrated in Walshe (1965), which presents force coefficients on a building for 10 different stages of erection, from open frames to partially clad to the then fully clad building. The wind load on the model when fully clad is less than that during several stages of erection.

C5.3.6.1 F_T and F_S

While the maximum wind load normal to the frame for a structure consisting of a single frame occurs when the wind direction is normal to the plane of the frame, this is not the case for a structure with multiple planes of frames. The maximum load normal to the plane of the frames occurs when the wind direction is typically 10° to 45° from the normal (Willford and Allsop 1990). This is because for oblique winds there is no direct shielding of successive columns and a larger area of frame is therefore exposed to the wind directly (without shielding) as the wind angle increases. Thus, the maximum wind load on one set of frames occurs at an angle that will also induce significant loads on the other set of frames (Willford and Allsop 1990, Georgiou et al. 1981).

Full and partial loading of structures given in ASCE 7-16 Section 27.3.5 was developed for clad structures only. The provisions of that section are not applicable to open-frame structures because of the different flow characteristics.

5.3.6.2 Shielding of Equipment
When, in the engineer's judgment, there is substantial shielding of

C5.3.6.2 Shielding of Equipment
This provision is similar to a provision in the Australian wind load standard,

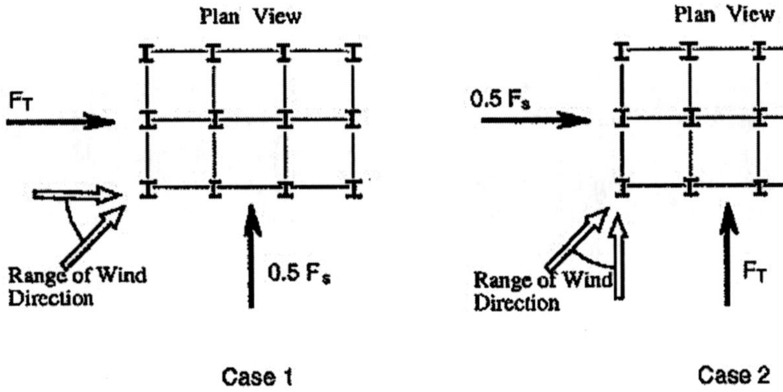

Notes:

1. F_s denotes the wind force on the structural frame and appurtenances in the indicated direction (excludes wind load on equipment, piping, and cable trays).
2. F_T denotes the total wind force on the structure in the direction indicated, which is the sum of forces on the structural frame and appurtenances, equipment, and piping.If appropriate, the equipment load may be reduced by considering shielding effects per Section 5.3.6.2.
3. Load combination factors applied to F_s may alternatively be determined by the detailed method of Appendix 5A and used in place of the 0.5 values shown.These values shall be calculated separately for Case 1 and Case 2.

Figure 5-2. Design load cases.

equipment by the structure or other equipment on a given level in the wind direction under consideration, the wind load on equipment in Section 5.3.6.1 on that level may be reduced by the shielding factor, η_{equip}

$$\eta_{equip} = \exp[-1.4\ (C_f\ \varepsilon)^{1.5}] \quad (5\text{-}5)$$

The force coefficient, C_f, is for the frame according to Section 5.3.3 or Appendix 5A. The solidity ratio, ε, is defined in Section 5.3.4. This factor is applied to the equipment force coefficient. The wind load on any equipment or portion thereof that extends above the top of the structure should not be reduced.

AS/NZS 1170.2:2011 (Standards Australia 2011), for reducing the wind load on cylindrical ancillaries located inside square-sectioned lattice towers. The provision in the Australian standard has its origin in data published by the Engineering Sciences Data Unit (ESDU) in the United Kingdom. Experimental work (Amoroso and Levitan 2009) suggests this provision is more accurate than the provision for equipment shielding given in the first edition of this guide.

5.3.6.3 Horizontal Torsion

Horizontal torsion (torsion about the vertical axis) may be a factor for open-frame structures. The engineer should consider the possibility of torsion in the design.

Consideration should be given to the application points of the wind load, especially in cases where the building framing is irregular and/or equipment locations are not symmetric.

C5.3.6.3 Horizontal Torsion

The line of action of the wind load may not coincide with the center of rigidity of the structure. In this case, the wind force may produce torsional loads on the structure.

5.4 Partially Clad Structures

This section is intended to address structures with cladding on less than four exterior walls. The wind loads on such structures vary considerably depending on the cladding arrangement and the wind direction. The forces calculated in this section are intended to be applied to the main wind-force-resisting system and are not intended for design of the components and cladding.

The different configurations of partially clad structures are illustrated in Figure 5-3. For all cases, forces along both axes should be applied simultaneously.

For open frames with only one vertical face having cladding, $C_f = 1.4$ for forces acting normal to the clad face. For forces acting parallel to the clad face, C_f shall be determined using the methods for open frames in Section 5.3 or Appendix 5A.

For open frames with cladding on two opposite, parallel faces, $C_f = 2.3$ for forces acting normal to the

C5.4 Partially Clad Structures

The force coefficients provided in this section are derived from wind tunnel tests on partially clad structures (Amoroso et al. 2010). These experiments were of limited parametric extent, and therefore the guide provisions should not be interpreted as being all-inclusive. A significant and relevant general finding of the study was that considerably higher force coefficients can occur for partially clad arrangements than for fully clad structures of the same envelope geometry. Furthermore, high force coefficients are often generated simultaneously for both orthogonal structural axes. These findings are represented in the provisions of this section.

When determining wind loads on open-frame structures, it is often advantageous to perform calculations for each of several individual stories along the structure height. This procedure takes advantage of the changes in the vertical velocity profile and accounts for changes in the geometric arrangement of the structure. Because of the high force coefficients associated with partially clad faces. For forces acting parallel to the clad faces, C_f shall be determined using the methods for open frames in Section 5.3 or Appendix 5A.

For open frames with cladding on two adjacent, perpendicular, vertical faces, $C_f = 2.0$ for forces along both structural axes when the unclad faces are positioned generally windward of the clad faces. When the clad faces are positioned windward of the unclad faces, $C_f = 1.5$ for forces acting along each structural axis.

clad structures, using this procedure is particularly important when the partial cladding does not extend over the full height of the structure.

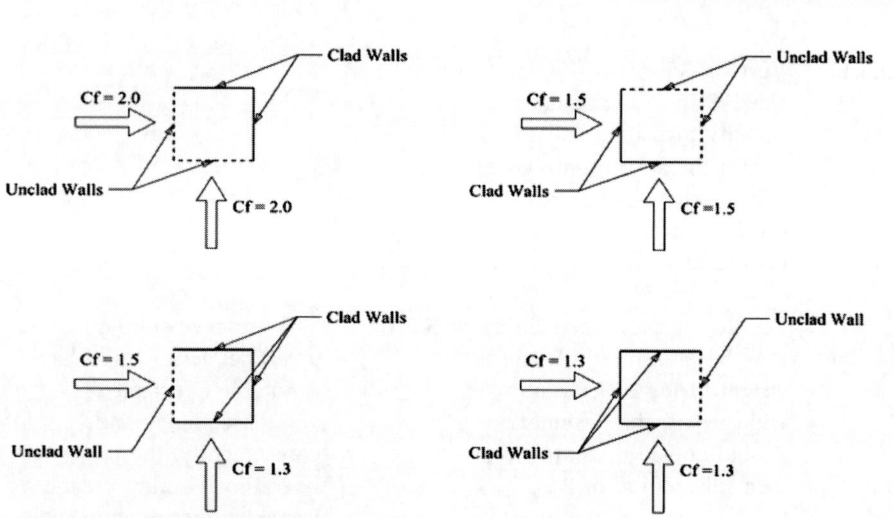

Figure 5-3. Configurations for partially clad structures.

For open frames with cladding on three vertical faces, $C_f = 1.5$ for forces acting normal to the unclad face when the unclad face is positioned on the windward side of the structure. When the unclad face is positioned on the leeward side of the structure, $C_f = 1.3$ for forces acting normal to the unclad face and $C_f = 1.3$ for forces acting along the axis parallel to the unclad face.

5.5 **Pressure Vessels**

Where vessel and piping diameters are specified, it is intended that insulation, if present, be included in the projected area. Insulation should not be included for stiffness when checking *H*/*D* for dynamic characteristics.

C5.5 **Pressure Vessels**

For tall slender vessels, vortex shedding may cause significant oscillating force in the crosswind direction. This means that the structure may experience significant loads in both the along-wind and crosswind directions at the same time. Crosswind forces such as vortex shedding are not addressed in this book.

5.5.1 **Vertical Vessels**

Use ASCE 7 to calculate velocity pressures and to obtain the appropriate gust effect factor, G_f, based on the governing empty or operating vessel frequency.

5.5.1.1 For those cases when fluid may be present inside the vessel during an extreme wind event, the designer should consider using the detailed method of ASCE 7 to calculate the gust effect

C5.5.1.1 The presence of fluid inside a vertical vessel could have a notable effect on the vessel frequency and, consequently, the total lateral load acting on the vessel because of wind. The factor, G_f, and consider both cases when the vessel is empty as well as full of fluid. The detailed method of ASCE 7 requires the engineer to determine the frequency of the vessel. The frequency can be determined by the following formulas:

$$f = 1/T(\text{s}) \quad (5\text{-}6)$$

detailed method of ASCE 7 provides guidance to calculate the gust effect factor, G_f. This procedure requires estimation of the vessel frequency, which can be accomplished using the equation for natural frequency.

A percentage of the empty weight of the vessel is usually added to the weight of the vessel to account for the weight of piping, which is not included in the initial weight of the empty or operating weights. This percentage increase will affect the vessel frequency. When calculating this percentage increase, the engineer should be careful to use an appropriate percentage increase based on the empty weight of the vessel compared to the additional weight of piping. A common percentage is usually around 10% of the empty weight of the vessel.

5.5.1.2 Simplified Method

If detailed information (e.g., number of platforms, platform size) is unknown at the time of design of the foundation/piles, the following approach may be used:

1. For the projected width, add 5 ft (1.52 m) to the diameter of the vessel, or add 3 ft (0.91 m) plus

C5.5.1.2 Simplified Method

1. With limited information on the vessel and appurtenances, this simple approach gives reasonably consistent results.

Natural period of vibration: Uniform vertical cylindrical steel vessel

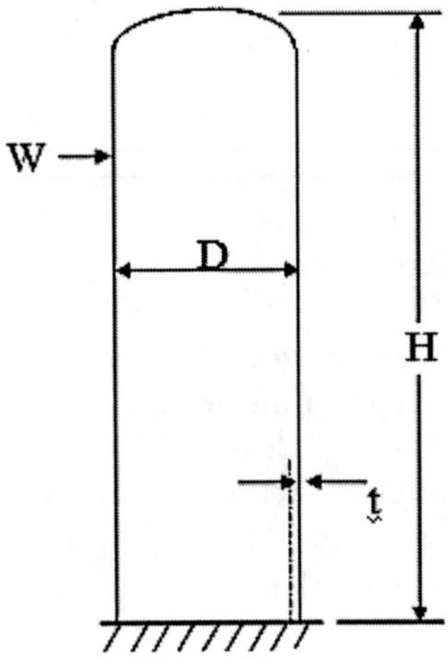

$$T = \frac{7.78}{10^6}\left(\frac{H}{D}\right)^2 \sqrt{\frac{12WD}{t}} \quad (5.7)$$

where:
T = Period (sec)
W = Weight (lb/ft)
H = Height (ft)
D = Diameter (ft)
t = Shell Thickness (inch)

Natural period of vibration: Nonuniform vertical cylindrical vessel

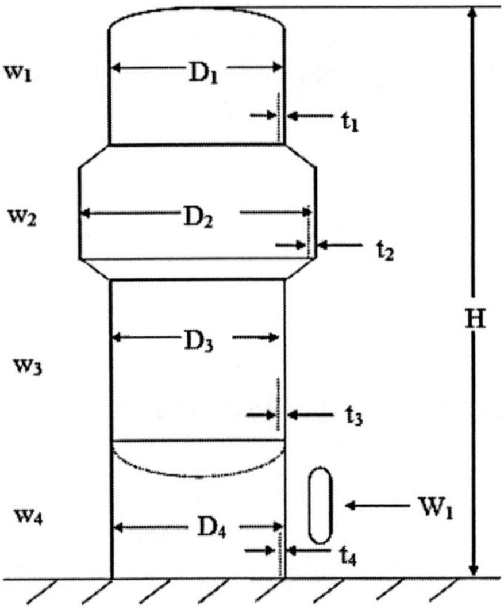

$$T = \left(\frac{H}{100}\right)^2 \cdot \sqrt{\frac{\Sigma w.\Delta\alpha + \frac{1}{H}.\Sigma W.\beta}{\Sigma E.D^3.t.\Delta\gamma}} \quad (5.8)$$

Where:

T = period (sec)
H = overall height (ft)
w = distributed weight (lbs/ft) of each section
W = Weight (lb) of each Concentrated Mass
D = diameter (ft) of each section
t = shell thickness (inch) of each section
E = modulus of elasticity (millions of psi)

α, β, and γ are coefficients for a given level depending on h_x/H ratio of the height of the level above grade to the overall height. $\Delta\alpha$ and $\Delta\gamma$ are the difference in the values of α and γ, from the top to the bottom of each section of uniform weight, diameter and thickness. β is determined and for each concentrated mass.

Note: Values of α, β, and γ are listed in Table 5-2.

Table 5-2. Coefficients for Determining Period of Vibration of Free-Standing Cylindrical Shells with Nonuniform Cross Section and Mass Distribution.

h_x/H	α	β	γ	h_x/H	α	β	γ
1.00	2.103	8.347	1.000000	0.50	0.1094	0.9863	0.95573
0.99	2.021	8.121	1.000000	0.49	0.0998	0.9210	0.95143
0.98	1.941	7.898	1.000000	0.48	0.0909	0.8584	0.94683
0.97	1.863	7.678	1.000000	0.47	0.0826	0.7987	0.94189
0.96	1.787	7.461	1.000000	0.46	0.0749	0.7418	0.93661
0.95	1.714	7.248	0.999999	0.45	0.0578	0.6876	0.93097
0.94	1.642	7.037	0.999998	0.44	0.0612	0.6361	0.92495
0.93	1.573	6.830	0.999997	0.43	0.0551	0.5372	0.91854
0.92	1.506	6.626	0.999994	0.42	0.0494	0.5409	0.91173
0.91	1.440	6.425	0.999989	0.41	0.0442	0.4971	0.90443
0.90	1.377	6.227	0.999982	0.40	0.0395	0.4557	0.89679
0.89	1.316	6.032	0.999971	0.39	0.0351	0.4167	0.88864
0.88	1.256	5.840	0.999956	0.38	0.0311	0.3801	0.88001
0.87	1.199	5.652	0.999934	0.37	0.0275	0.3456	0.87033
0.86	1.143	5.467	0.999905	0.36	0.0242	0.3134	0.86123
0.85	1.090	5.285	0.999867	0.35	0.0212	0.2833	0.85105
0.84	1.038	5.106	0.999817	0.34	0.0185	0.2552	0. 84032
0.83	0.938	4.930	0.999754	0.33	0.0161	0.2291	0.82901
0.82	0.939	4.758	0.999674	0.32	0.0140	0.2050	0.81710
0.81	0.892	4.589	0.999576	0.31	0.0120	0.1826	0.80459
0.80	0.847	4.424	0.999455	0.30	0.010293	0.16200	0.7914
0.79	0.804	4.261	0.999309	0.29	0.008769	0.14308	0.7776
0.78	0.762	4.102	0.999133	0.28	0.007426	0.12576	0.7632
0.77	0.722	3.946	0.998923	0.27	0.006249	0.10997	0.7480
0.76	0.683	3.794	0.998676	0.26	0.005222	0.09564	0.7321
0.75	0.646	3.645	0.998385	0.25	0.004332	0.08267	0.7155
0.74	0.610	3.499	0.998047	0.24	0.003564	0.07101	0.6981
0.73	0.576	3.356	0.997656	0.23	0.002907	0.06056	0.6800
0.72	0.543	3.217	0.997205	0.22	0.002349	0.05126	0.6610
0.71	0.512	3.081	0.996689	0.21	0.001878	0.04303	0.6413
0.70	0.481	2.949	0.996101	0.20	0.001485	0.03579	0.6207
0.69	0.453	2.820	0.995434	0.19	0.001159	0.02948	0.5902
0.68	0.425	2.694	0.994681	0.18	0.000893	0.02400	0.5769
0.67	0.399	2.571	0.993834	0.17	0.000677	0.01931	0.5536
0.66	0.374	2.452	0.992885	0.16	0.000504	0.01531	0.5295

(Continued)

Table 5-2. Coefficients for Determining Period of Vibration of Free-Standing Cylindrical Shells with Nonuniform Cross Section and Mass Distribution. (Continued)

h_x/H	α	β	γ	h_x/H	α	β	γ
0.65	0.3497	2.3365	0.99183	0.15	0.000368	0.01196	0.5044
0.64	0.3269	2.2240	0.99065	0.14	0.000263	0.00917	0.4783
0.63	0.3052	2.1148	0.98934	0.13	0.000183	0.00689	0.4512
0.62	0.2846	2.0089	0.98739	0.12	0.000124	0.00506	0.4231
0.61	0.2650	1.9062	0.98630	0.11	0.000081	0.00361	0.3940
0.60	0.2464	1.8068	0.98455	0.10	0.000051	0.00249	0.3639
0.59	0.2288	1.7107	0.98262	0.09	0.000030	0.00165	0.3327
0.58	0.2122	1.6177	0.98052	0.08	0.000017	0.00104	0.3003
0.57	0.1965	1.5279	0.97823	0.07	0.000009	0.00062	0.2669
0.56	0.1816	1.4413	0.97573	0.06	0.000004	0.00034	0.2323
0.55	0.1676	1.3579	0.97301	0.05	0.000002	0.00016	0.1965
0.54	1.1545	1.2775	0.97007	0.04	0.000001	0.00007	0.1597
0.53	0.1421	1.2002	0.96683	0.03	0.000000	0.00002	0.1216
0.52	0.1305	1.1259	0.96344	0.02	0.000000	0.00000	0.0823
0.51	0.1196	1.0547	0.95973	0.01	0.000000	0.00000	0.0418

the diameter of the largest pipe to the diameter of the vessel, whichever is greater. This will account for platforms, ladders, nozzles, and piping below the top tangent line.

2. The vessel height should be increased by one vessel diameter to account for a large-diameter pipe and platform attached above the top tangent, as is the case with most tower arrangements.
3. The increases in vessel height or diameter to account for wind on

2. This method is an approximation to alleviate the need for some rather tedious calculations based on gross assumptions.

3. As noted in Section 5.5, insulation should not be included in the *H*/*D* calculations.

appurtenances should not be used in calculating the *H*/*D* ratio for force coefficients or flexibility.

4. The force coefficient (C_f) should be determined from ASCE 7 Figure 29.4-1, assuming a rough surface.

4. Based on wind tunnel studies conducted by Amoroso (2007), the force was underestimated using the simplified method from the previous edition of this guideline. Therefore, a rough surface should be assumed, which, in the absence of more detailed design information, will likely result in a more appropriate (i.e., higher) estimate of the total wind load. If ribbed insulation will be used, then D'/D should be calculated.

5. Use the wind directionality factor, K_d, of 0.95.

5. Vertical vessels typically include attached piping, platforms, ladders, and other appurtenances that result in a non-axisymmetric structural system.

5.5.1.3 Detail Method

If most design detail items (e.g., platforms, piping, ladders) of the vessel are known, the following method should be used:

1. For the projected width, add 1.5 ft (0.46 m) to the vessel diameter to account for ladders, nozzles, and piping 8 in. (0.2 m) or smaller and add the diameter of the

C5.5.1.3 Detail Method

This method will provide more accurate values for foundation design.

largest line coming from the top portion of the vessel.

2. The force coefficient (C_f) should be taken from ASCE 7, Figure 29.4-1, based on appropriate roughness at the vessel surface.

3. For pipes outside the projected width of the vessel (defined in Step 1) larger than 8 in. (0.2 m), including insulation, use the projected area of the pipe and a force coefficient (C_f) of 0.7.

   For pipes inside the projected width of the vessel (defined in Step 1) larger than 8 in. (0.2 m), including insulation, and more than five pipe diameters from the vessel surface, add the projected area of the pipe and use a force coefficient (C_f) of 0.7.

3. For a moderately smooth surface, C_f is determined from ASCE 7, Figure 29.4-1.

4. For platforms, use the projected area of the support steel and a force coefficient (C_f) of 2.0.

   For handrails, use the values for area and force coefficient from Table 5-1.

   Where the railing extends beyond the

4. The front and back systems of railings are far enough apart to preclude shielding.

vessel, the projected area of two sets of railing systems for the projection beyond the vessel should be used.

5. Use the wind directionality factor, K_d, of 1.0, for the vessel only.

   The K_d values for the piping, platforms, ladders, and other appurtenances attached to the vessel, and structural supports for the vessel should be taken from ASCE 7-16 Table 26.6-1.

5. The vertical vessel alone exhibits radial symmetry in the plan view. Therefore, K_d of the vessel alone is determined from ASCE 7, Table 26.6-1 for a round structure similar to chimneys and tanks. Per ASCE 7, Table 26.6-1, $K_d = 1.0$ can be used for the piping and $K_d = 0.85$ can be used for the platforms, ladders, and other appurtenances attached to the vessel.

5.5.1.4 Closely Spaced Vertical Vessels

When a vertical vessel is located at a center-to-center distance of three diameters or less (based on the smaller of the two vessels) from a nearby vertical vessel, an increase of 20% is to be applied to the force coefficient value in the absence of model wind tunnel testing or full-scale data. This increase is to be applied for wind directions oriented approximately perpendicular to the axis connecting the center lines of the vessels.

C5.5.1.4 Closely Spaced Vertical Vessels

Research has shown that wind loads for closely spaced cylinders and similar structures can be amplified because of aerodynamic interference effects. This provision and the provision immediately following are modeled after a similar provision in ASME STS-1-2016 (ASME 2017) for the wind loading of steel stacks.

The vessel manufacturer should be advised of the increased wind load for closely spaced vessels.

5.5.1.5 Vertical Vessel with Large-diameter Piping
When a large-diameter pipe (greater than 8 in.) is located at center-to-center distance of three pipe diameters from the surface of a vertical vessel, the force coefficient applied to the pipe is to be increased by 20% in the absence of model wind tunnel testing or full-scale data. This increase is to be applied for wind directions oriented approximately perpendicular to the axis connecting the center lines of the pipe and the vessel.

5.5.2 Horizontal Vessels

5.5.2.1 No check for dynamic properties is required.

5.5.2.2 For the projected diameter, add 1.5 ft (0.46 m) to the insulated diameter to account for ladders, nozzles, and pipes 8 in. (0.2 m) (including insulation) or smaller.

5.5.2.3 For wind perpendicular to the long axis of the vessel (transverse wind), the force coefficient (C_f) should be determined from ASCE 7, Figure 29.4-1.

C5.5.2.3 Use B/D (length/diameter) to determine C_f.

5.5.2.4 For wind in the longitudinal direction, use C_f of 0.5 for a rounded head and 1.2 for a flat head.

C5.5.2.4 This wind direction will seldom control design of foundations.

5.4.2.5 For a pipe larger than 8 in. (0.2 m), including insulation, use the projected area of the pipe and a force coefficient (C_f) of 0.7.

5.5.2.6 For platforms, use the projected area of the support steel and a force coefficient (C_f) of 2.0.

For handrails, use the values for area and force coefficient from Table 5-1.

Use the projected area of each railing system.

C5.5.2.6 The reason for projecting the front and back railing system is that they are far enough apart to preclude shielding.

5.5.2.7 For supports, use the actual projected area. The value of C_f should be 1.3 for concrete pedestals. For steel supports, use the method described for platforms.

C5.5.2.7 The 1.3 factor is used because a pedestal is similar to a bluff rectangular body.

5.5.2.8 Use the wind directionality factor, K_d, of 0.85.

C5.5.2.8 Unlike vertical vessels and spheres, horizontal vessels do not have radial symmetry in the plan view.

5.5.3 Spheres

5.5.3.1 No check for dynamic properties is required.

5.5.3.2 For the projected diameter, add 1.5 ft (0.46 m) to the insulated diameter to account for ladders, nozzles, and pipes 8 in. (0.2 m) (including insulation) or smaller.

5.5.3.3 Use $C_f = 0.5$ (for vessel only). Supports should be evaluated separately.

C5.5.3.3 ASCE 7 does not provide C_f values for spheres. Daugherty and Franzini (1977) provide drag coefficients for a sphere as a function of Reynolds number (*Re*). For *Re* in the range of 10^5, the value is approximately equal to 0.5.

5.5.3.4 For a pipe larger than 8 in. (0.2 m), including insulation, use the projected area of the pipe and use a force coefficient (C_f) of 0.7.

C5.5.3.4 The value of C_f is determined from ASCE 7, Figure 29.4-1 for a moderately smooth surface.

5.5.3.5 For platforms, use the projected area for the support steel and a force coefficient (C_f) of 2.0.

For handrails, use the values for area and force coefficient from Table 5-1. Use the projected area of each railing system.

C5.5.3.5 The reason for projecting the front and back railing system is that they are far enough apart to preclude shielding.

5.5.3.6 For supports, use the actual projected area. The value of C_f should be 1.3 for rectangular concrete columns and 0.7 for circular columns. For steel supports, use the method described for platforms.

5.5.3.7 Use the wind directionality factor, K_d, of 1.0.

C5.5.3.7 Spheres have radial symmetry in the plan view.

5.6 Cooling Towers

Use the envelope procedure of ASCE 7, Chapter 28 to calculate wind pressures on the MWFRS and use the

C5.6 Cooling Towers

The Cooling Technology Institute (CTI) standard CTI Technical Paper TP06-22 (2006) cites ASCE 7 as the design basis for calculating procedure in Chapter 30 to calculate wind pressure on components and cladding.

If located on top of a building or other supporting structure in a manner where the top of the tower is at a height greater than 60 ft, use the directional procedure of ASCE 7.

For the fan cylinder (or shroud), calculate wind pressures across its surface separately using the following equation:

$$p_z = K_z\, K_{zt}\, (V/28)^2\, I \qquad (5\text{-}9)$$

where p_z is in pounds per square foot, and V is in miles per hour

$$F_z = p_z\, H\, D \qquad (5\text{-}10)$$

where
F_z is in pounds,
H = height (ft), and
D = diameter (ft).

wind loads on wooden cooling towers.

The cylinder is similar in geometry and aspect ratio to open top tanks. The American Petroleum Institute (API) provides design specifications for such structures. The equation provided here was derived based on API 650 and ASCE 7.

5.7 Air-Cooled Heat Exchanger (Air Coolers or Fin-Fans)

In Table 5-3, the force coefficients C_f for isolated solid blocks are provided to calculate wind loads on air coolers. The effective area A_e equals b times c.

C5.7 Air-Cooled Heat Exchanger (Air Coolers or Fin-Fans)

The refinery and petrochemical industries have often installed air coolers as cooling facilities. At one time, a specific manufacturer called its air coolers "fin-fans," and the term later came to mean any air cooler. For the convenience of piping access and conservation of plot space, the most common location to install air coolers is at the top of new or existing pipe racks. Most air coolers have maintenance access platforms.

For the estimation of wind loads, air coolers have been treated similarly to buildings at grade. However, because air coolers are elevated, it is the opinion of this committee that they may be considered as isolated blocks that allow air to flow beneath them as well as over and around them.

ASCE 7 does not have force coefficients for solid blocks in open space. In the past, engineers used their own engineering judgment to calculate wind loads on air coolers.

Multiple cooler units can be treated as one larger body for wind load estimation purposes when the clear spacing between units in the along-wind direction is less than or equal to one-half the crosswind dimension b of the individual unit. In this case, the a dimension of the combined units is the sum of the a dimension for each unit plus the clear space.

This provision applies only to the wind load on the air cooler unit. The dimensions should not include platforms and other appurtenances outside the outline of the air cooler unit.

5.8 Compressor Shelters

This section is intended to clarify the wind loads for compressor shelters that have a roof and four exterior walls, with the opening area in each wall being less than 80% but typically more than 20% of the gross area of the wall.

For wind design, compressor shelters are typically treated as buildings per ASCE 7. The compressor shelters as described cannot be classified as open buildings or enclosed buildings per Chapter 26 and Table 26.13-1 of the standard. These types of compressor shelters are usually classified as partially open buildings instead of partially enclosed buildings per ASCE 7-16. As a result, the internal pressure coefficients for a partially open building (i.e., ±0.18 for GC_{pi}) as shown in Table 26.13-1 of the standard, should be used for these types of compressor shelters.

C5.8 Compressor Shelters

For wind design, ASCE 7 has four enclosure classifications for buildings: enclosed buildings, partially enclosed buildings, partially open buildings, and open buildings.

Compressor shelters consisting of a roof and four exterior walls with the opening area in each wall less than 80% but typically more than 20% of the gross area of the wall are usually classified as partially open buildings per Chapter 26 and Table 26.13-1 of ASCE 7 The internal pressure coefficients for partially open buildings are the same as those for enclosed buildings (i.e., ±0.18 for GC_{pi}).

Prior to the 2016 edition of ASCE 7, there were only three enclosure classifications for the wind design of buildings: enclosed buildings, partially enclosed buildings, and open buildings. Some engineers have treated these types of compressor shelters as partially enclosed buildings with internal pressure coefficients of ±0.55 for GC_{pi}. From the perspective of the internal pressure coefficient, it is incorrect, but usually conservative, to treat these types of compressor shelters as partially enclosed buildings.

Table 5-3. C_f for Isolated Blocks in Datum Conditions.

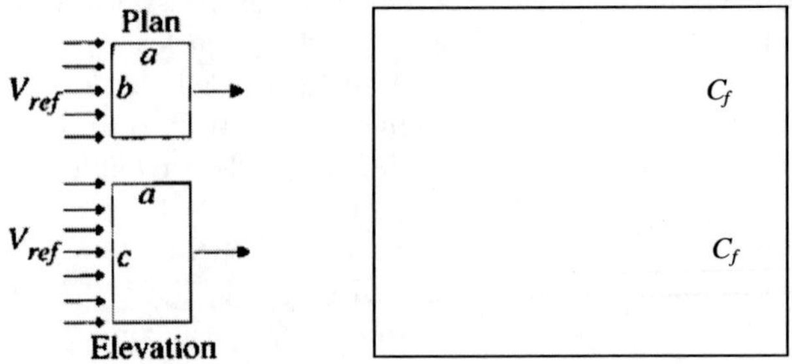

		a/*b*								Max C_f	
		0.00	0.25	0.50	1.00	1.50	2.00	2.50	3.00	C_f	*a*/*b*
	0.25	1.20	1.20	1.00	0.92	0.91	0.91	0.91	0.90	1.23	0.20
	0.50	1.16	1.21	1.16	0.97	0.92	0.91	0.91	0.90	1.21	0.34
	1.00	1.14	1.15	1.18	1.16	1.02	0.95	0.92	0.90	1.20	0.70
c/*b*	2.00	1.16	1.17	1.19	1.17	1.04	0.97	0.93	0.92	1.21	0.70
	5.00	1.21	1.22	1.24	1.22	1.11	1.03	0.99	0.97	1.27	0.70
	8.00	1.28	1.29	1.35	1.30	1.17	1.09	1.04	1.01	1.36	0.70
	10.00	1.31	1.34	1.41	1.36	1.24	1.14	1.08	1.04	1.46	0.70

Source: Generated from Figure 1 of ESDU (1978).

APPENDIX 5A. ALTERNATE METHOD FOR DETERMINING C_f AND LOAD COMBINATIONS FOR OPEN-FRAME STRUCTURES

5A.1 Background

Maximum wind force normal to the face of a rectangular enclosed building occurs when the wind direction is normal to the building face. The same is true for wind load on a single frame or solid sign. However, this is not the case for an open-frame structure with more than one frame. As the wind direction moves away from the normal and more toward a quartering wind, columns that once lined up neatly behind each other, shielding each other, become staggered and exposed to the full wind. Additionally, the area of the structure projected on a plane normal to the wind also increases.

The variation of the wind loads along each principal axis of a rectangular open-frame structure with the direction of wind is shown in Figure 5A-1, and for the structure and wind angle of attack defined in Figure 5A-2. It can readily be seen that when one frame set experiences its maximum frame load *A* or *D*, the frame set along the other axis experiences a wind force *C* or *B*, respectively, thus the need for the load combinations of Section 5.3.6.1. In those provisions, the load at *C* is roughly estimated to be 50% of *A*, and *B* is estimated to be 50% of the load at *D*. In actuality, the loads on the secondary axis can range from about 25% to 75% of the primary axis load, depending on many factors including spacing ratio, number of frames, and solidity ratio. This appendix provides a method to obtain a better estimate of the simultaneously acting load on the secondary axis.

5A.2 Force Coefficients

This method provides force coefficients C_f for a greater range of ε and S_F/B values than the method of Section 5.3.3, as well as providing an estimate of α_{max}. For cases where both methods are applicable, the methods will generally yield very similar results. Nadeem (1995) and Nadeem and Levitan (1995, 1997) discuss this method in greater detail. The procedure is as follows:

1. Determine ε, S_F/B, and N for the principal axis under consideration as per Section 5.3.4 and Figure 5-1.
2. Estimate the wind angle of attack that maximizes the force parallel to the axis under consideration

$$\alpha_{max} = (10 + 58\varepsilon)^{o} \quad \text{for } 3 \leq N \leq 5 \qquad (5A\text{-}1)$$
$$\alpha_{max} = (16 + 52\varepsilon)^{o} \quad \text{for } 6 \leq N \leq 10 \qquad (5A\text{-}2)$$

3. Estimate the force coefficient C_f from Figure 5A-3 by the following procedure:
   (a) Determine C_f from Figure 5A-3(a) for wind angle of attack equal to α_{max} and appropriate spacing ratio S_F/B. This C_f is for a structure with $N = 3$ frames and a solidity ratio of $\varepsilon = 0.1$.
   (b) Determine C_f from Figure 5A-3(b) for a structure with $N = 3$ frames and a solidity ratio of $\varepsilon = 0.5$.
   (c) Interpolate between results of Steps a and b for the actual solidity ratio, yielding a force coefficient for the correct spacing and solidity ratios, and $N = 3$ frames, $C_{f,N=3}$.
   (d) Determine C_f from Figure 5A-3(c) for a structure with $N = 10$ frames and a solidity ratio of $\varepsilon = 0.1$.
   (e) Determine C_f from Figure 5A-3(d) for a structure with $N = 10$ frames and a solidity ratio of $\varepsilon = 0.5$.

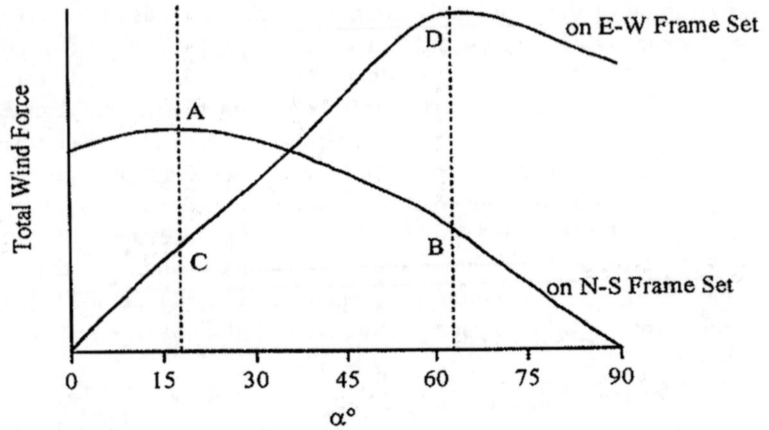

Figure 5A-1. Variation of wind load versus wind direction.

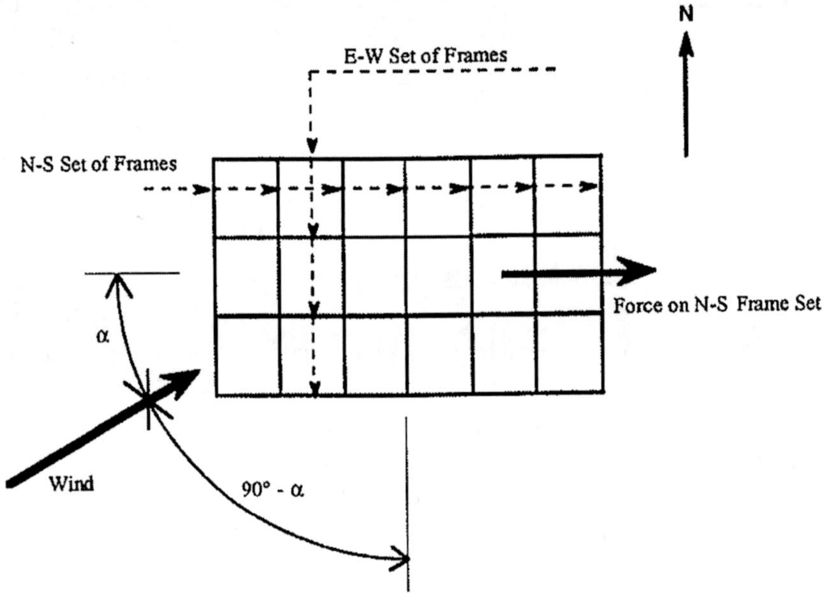

Figure 5A-2. Plan view of structural framing.

(f) Interpolate between results of Steps d and e for the actual solidity ratio, yielding a force coefficient for the correct spacing and solidity ratios, and $N = 10$ frames, $C_{f,N=10}$.

(g) Determine C_f for the axis under consideration by interpolating between $C_{f,N=3}$ and $C_{f,N=10}$ for the actual number of frames.

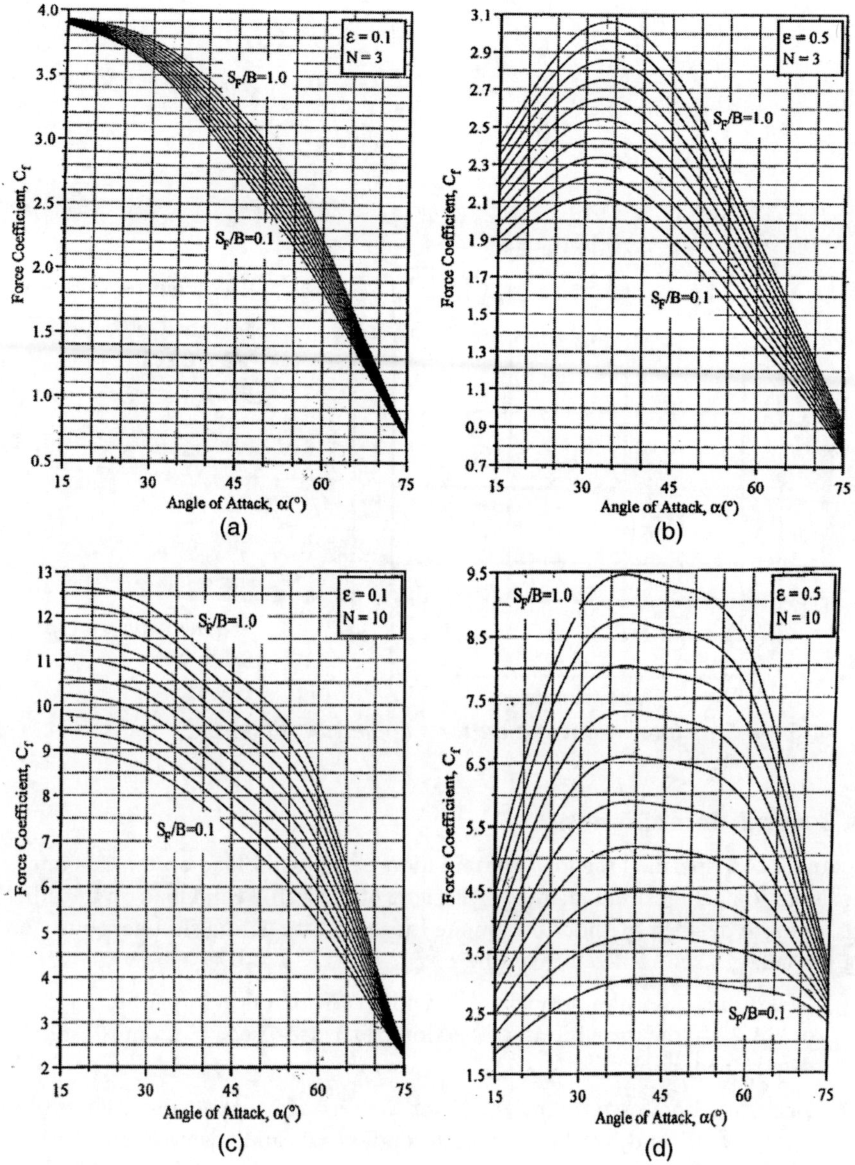

Figure 5A-3. Force coefficients, C_f.

If the structure has exactly 3 or 10 frames, only Steps a–c or d–f, respectively, need to be used. Similarly, if a structure has a solidity ratio very near 0.1 or 0.5, only one interpolation between Figures 5A-3(a and c) or 5A-3(b and d), respectively, would be necessary.

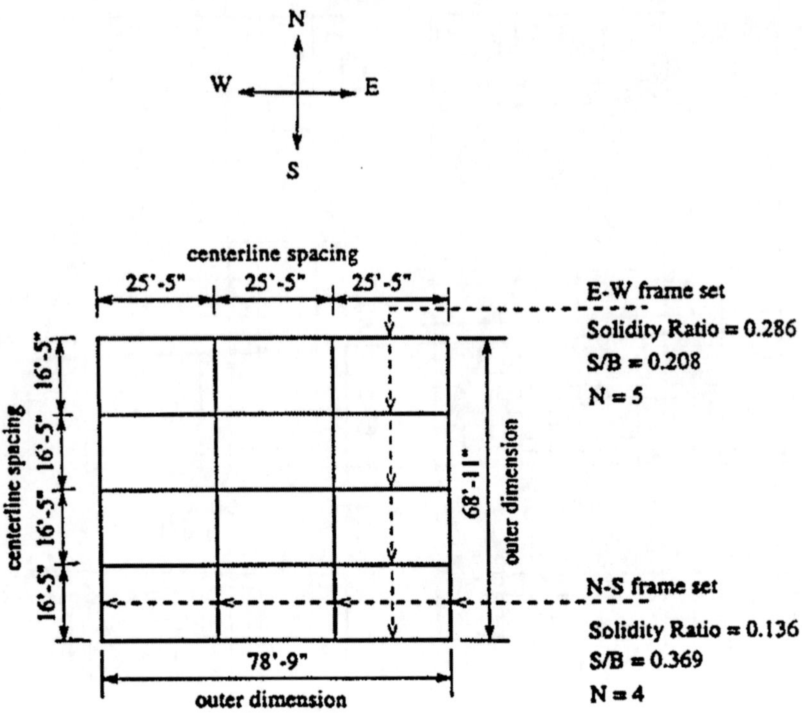

Figure 5A-4. Structural framing plan for sample calculations.

5A.3 Load Combinations

Section 5.3.6.1 specified the load combination of full wind load on the axis under consideration acting simultaneously with 50% of the frame wind load on the other axis. A more detailed method to estimate the wind load acting simultaneously on the secondary axis frames is given here:

1. Determine C_f for the principal axis under consideration as per Section 5.3.3 or 5A.2. If the provisions of Section 5.3.3 are used, α_{max} must still be determined as per Section 5A.2.
2. Determine the force coefficient C_f for the secondary axis from Figure 5A-3 using ε, S_F/B, and N values for the secondary axis and a wind angle of attack of $(90° - \alpha_{max})$. Step 3 in Section 5A.2 explains how to obtain C_f from Figure 5A-3.

5A.4 Sample Calculations

The proposed method has been used to calculate the force coefficients for a structure whose plan is shown in Figure 5A-4. The important frame set properties can be summarized as follows:

For winds nominally from west to east [i.e., winds normal to the north–south (N-S) frame set]

$\varepsilon = 0.136$
$N = 4$
$S_F = 25.4$ ft (7.75 m)
$B = 68.9$ ft (21.0 m)
$S_F/B = 25.4/68.9 = 0.369$

For winds nominally from south to north [i.e., winds normal to the east–west (E-W) frame set]

$\varepsilon = 0.286$
$N = 5$
$S_F = 16.4$ ft (5.0 m)
$B = 78.75$ ft (24.0 m)
$S_F/B = 16.4/78.75 = 0.208$

5A.4.1 Determining Force Coefficients

To determine the force coefficient for the E-W structural axis (winds nominally normal to the N-S frame), first estimate the wind angle of attack at which this maximum load will occur. Because $N = 4$ and $\varepsilon = 0.136$ for the N-S frame set, Equation (5A-1) yields

$$\alpha_{max} = 10 + 58\ (0.136) = 18°$$

From Figures 5A-3(a and b), for $S_F/B = 0.369$, $C_f = 3.87$ and 2.10 for structures with $\varepsilon = 0.1$ and $\varepsilon = 0.5$, respectively, and $N = 3$ frames. Interpolating between these two values for $\varepsilon = 0.136$

$$C_{f,N=3} = 3.87 - [(3.87 - 2.10)\ /\ (0.5 - 0.1)]\ (0.136 - 0.1) = 3.71$$

From Figures 5A-3(c and d), for $S_F/B = 0.369$, $C_f = 10.08$ and 3.15 for structures with $\varepsilon = 0.1$ and $\varepsilon = 0.5$, respectively, and $N = 10$ frames. Interpolating between these two values for $\varepsilon = 0.136$

$$C_{f,N=10} = 10.08 - [(10.08 - 3.15)\ /\ (0.5 - 0.1)]\ (0.136 - 0.1) = 9.46$$

Interpolating between the two previous results of $C_{f,N=3} = 3.71$ and $C_{f,N=10} = 9.46$ for the case $N = 4$

$$C_f = 3.71 + [(9.46 - 3.71)\ /\ (10 - 3)]\ (4 - 3) = 4.53$$

gives the maximum force coefficient for the N-S set of frames, occurring near $\alpha_{max} = 18°$.

5A.4.2 Determining Load Combinations

While the maximum wind load is acting on the N-S set of frames, the wind simultaneously acts on the E-W set of frames at angle of attack of 90° – 18° = 72°.

The force coefficient for the E-W frames is determined as per Step 2 of Section 5A.3.

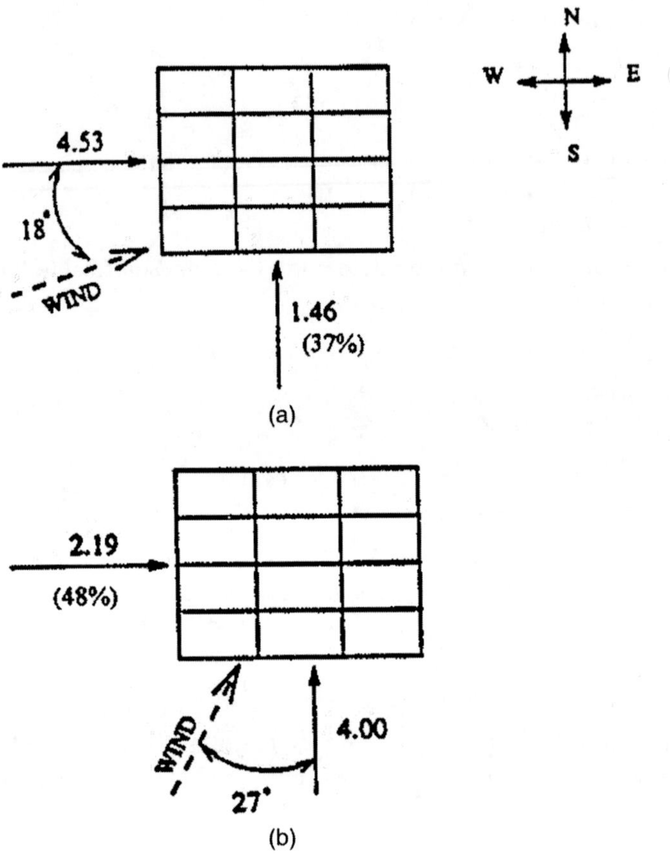

Figure 5A-5. Force coefficients, C_f, for example design load combinations (a) considering maximum load on N-S frame set, and (b) considering maximum load on E-W frame set.

For $\alpha = 72°$, $S_F/B = 0.208$, and $\varepsilon = 0.286$, interpolation between Figures 5A-3(a and b) yields $C_{f,N=3} = 0.91$. Interpolation between Figures 5A-3(c and d) yields $C_{f,N=10} = 2.84$.

One more interpolation between $C_{f,N} = 3$ and $C_{f,N} = 10$ for $N = 5$ frames yields $C_f = 1.46$, which is the force coefficient for the wind load acting on the E-W frame set while the N-S set is experiencing its maximum wind load. This combination of loads is shown in Figure 5A-5(a).

This entire procedure should now be repeated assuming that maximum wind load acts on the E-W set of frames. For this case, $\alpha_{max} = 27°$, $C_f = 4.0$ for the E-W frames, with a simultaneously acting load of $C_f = 2.19$ for the N-S frame set, as shown in Figure 5A-5(b).

For the case of full wind load on the E-W axis, use of this alternate procedure reduced the wind load acting simultaneously on the N-S axis from 50% (Section

5.3.6.2) to 1.46/4.00 = 37%. The load combination for full wind on the N-S structure axis remained close to the recommended 50% at 2.19/4.53 = 48%.

APPENDIX 5B. HIGH-SOLIDITY OPEN-FRAME STRUCTURES

5B.1 Background

The arrangement and congestion of structural members, equipment, piping, electrical items, and appurtenances such as ladders and cages, handrails, and stairs in many open-frame process structures creates structures with solidity ratios in excess of those that are the basis for the recommendations in Section 5.3 and Appendix 5A. Figure 5B-1 provides a typical example of a structure with a dense equipment and piping arrangement. An example of a more open structure for which Section 5.3 and Appendix 5A would be applicable is shown in Figure 5B-2.

5B.2 Force Coefficients

As the overall solidity of an open-frame structure and equipment increases, not only do wind load estimation methods become more cumbersome, but their accuracy may be diminished as the simple, empirical basis for the aerodynamic interaction effects among the framing and equipment becomes insufficient. It is advantageous to work toward a general understanding of the wind loads for high-solidity open-frame process structures because accounting for the effects of all the individual components becomes tedious and less accurate as the structures become more densely occupied. As a result of research performed at Louisiana State University, an analytical model for the force coefficient was derived for "wind-porous" structures. According to this model, the maximum force coefficient depends on the length-to-width aspect ratio and the solidity ratio. Equation (5B-1) expresses the maximum force coefficient for a densely occupied open-frame structure

$$C_{max} = \frac{C_0 \cdot \phi \cdot \sqrt{B^2 + L^2}}{2 \cdot B \cdot \left(1 + \frac{B^2}{L^2} - \frac{B}{L^2} \cdot \sqrt{B^2 + L^2}\right)} \qquad \text{(5B-1)}$$

where

C_0 = Equivalent solid body force coefficient,
ϕ = Total solidity ratio (including equipment),
B = Width of the structure, and
L = Length of the structure.

Comparisons of the analytical model with wind tunnel measurements for models consisting of multiple parallel lattice frames and fully three-dimensional frameworks showed good agreement in the trends, a low bias, and some scatter.

Figure 5B-1. Typical process structure with dense arrangement of equipment and piping.

This is illustrated in Figure 5B-3. The experimental data displayed the increase in force coefficient with the plan aspect ratio, which the analytical model predicts. It is possible to describe the upper-limit force coefficient only as a function of the length-to-width ratio. This requires a selection of the empirical parameter, C_0, which ensures the results from the analytical model adequately envelope the force coefficient for porous structures. A choice of $C_0 = 1.4$ met this criterion for the experimental data examined in the study.

To provide a method for estimating an envelope value for the force coefficient for open-frame structures, it is proposed that the following relationships be allowed for calculating the gross force coefficients for open-frame structures with total projected solidity ratios greater than 50% (including all framing, equipment, vessels, piping, and other appurtenant structures):

Figure 5B-2. Typical process structure with open equipment and piping arrangement.

$$
\begin{aligned}
C_f &= \frac{1}{4} \cdot \left(\frac{L}{B}\right)^2 + 1.4 \quad \text{for } L/B < 1.5 \\
C_f &= \frac{2}{3} \cdot \left(\frac{L}{B}\right) + 0.9 \quad \text{for } L/B \geq 1.5
\end{aligned}
\tag{5B-2}
$$

where L and B are plan dimensions corresponding to the along-wind length and the across-wind width, respectively. The force coefficient is to be applied using the gross area of the structure, A_g.

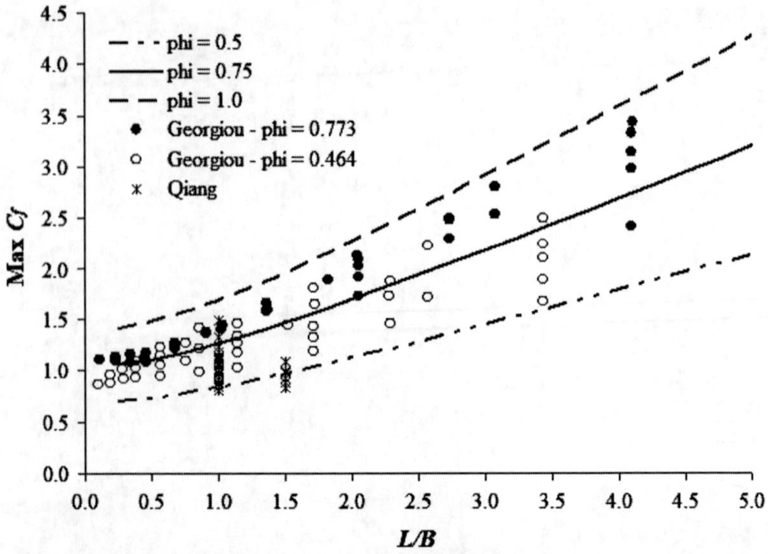

Figure 5B-3. Maximum C_f with respect to plan aspect ratio. The lines indicate the results of Equation (5B-1), and the points represent wind tunnel measurements.
Source: Amoroso and Levitan (2011).

5B.3 Load Combinations

For this method, the same load cases in Section 5.3.6 can be used with the exception that F_S should be replaced by F_T because F_S is not calculated separately in this method. The intent of this method is to facilitate a reasonably conservative design of the structural frame without requiring detailed knowledge of the equipment, which would also allow the structure to be designed to safely accommodate future process changes.

References

Amoroso, S. 2007. "Wind loads for petrochemical structures." Ph.D. dissertation, Dept. of Civil and Environmental Engineering, Louisiana State Univ.

Amoroso, S., K. Hebert, and M. Levitan. 2010. "Wind tunnel tests for mean wind loads on partially clad structures." *J. Wind Eng. Ind. Aerodyn.* **98** (12): 689–700.

Amoroso, S., and M. Levitan. 2009. "Recent research for wind loads on petrochemical structures." In *Proc., ASCE/SEI 2009 Structures Congress*, Austin, TX, 2088–2097.

Amoroso, S., and M. Levitan. 2011. "Wind loads for high-solidity open-frame structures." *Wind Struct.* **14** (1): 1–14.

API (American Petroleum Institute). 2013. *Welded steel tanks for oil storage.* 12th ed. Washington, DC: API.

ASCE. 2017. *Minimum design loads and associated criteria for buildings and other structures.* ASCE 7-16. Reston, VA: ASCE.

ASME. 2017. *Steel stacks.* ASME STS-1-2016. New York: ASME.

Cook, N. J. 1990. *The designer's guide to wind loading of building structures part 2: Static structures*. London: Butterworths.

CTI (Cooling Technology Institute). 2006. *Wind load rated packaged cooling towers*. CTI TP06-22. Houston: CTI.

Daugherty, R. L., and J. B. Franzini. 1977. *Fluid mechanics with engineering applications*. New York: McGraw-Hill.

ESDU (Engineering Sciences Data Unit). 1978. *Fluid forces, pressures and moments on rectangular blocks*. ESDU 71016. London: ESDU.

Georgiou, P. N. 1979. "Wind loads on building frames." M.E. Sc. thesis, Dept. of Civil and Environmental Engineering, Univ. of Western Ontario.

Georgiou, P. N., B. J. Vickery, and R. Church. 1981. "Wind loading on open framed structures." In Vol. **1** of *Proc., Program and Workshop Notes, CWOWE III: 3rd Canadian Workshop on Wind Engineering*, Vancouver, BC, Canada, 1–19.

Levitan, M., L. Qiang, and S. Amoroso. 2004. "Wind tunnel tests on open-frame industrial/petrochemical structures." In *Proc., 5th Int. Colloquium on Bluff Body Aerodynamics and Applications*, Ottawa.

Nadeem, A. 1995. "Wind loads on open frame structures." M.S. thesis, Dept. of Civil and Environmental Engineering, Louisiana State Univ.

Nadeem, A., and M. L. Levitan. 1995. "A refined method for calculating wind loads on open frame structures." In *Proc., 9th Int. Conf. on Wind Engineering*, New Delhi, India.

Nadeem, A., and M. Levitan. 1997. "A refined method for calculating wind load combinations on open-framed structures." *J. Wind Eng. Ind. Aerodyn.* **72** (Nov–Dec): 445–453.

Qiang, L. 1998. "Wind tunnel tests for wind loads on open frame petrochemical structures." Master's thesis, Dept. of Civil and Environmental Engineering, Louisiana State Univ.

Standards Australia. 2011. *Structural design actions—Part 2: Wind actions*. AS/NZS 1170.2:2011. Sydney, Australia: Standards Australia.

Walshe, D. E. 1965. *Measurements of wind force on a model of a power station boiler house at various stages of erection*. NPL Aero Rep. No. 1165. Teddington, UK: National Physical Laboratory, Aerodynamics Division.

Whitbread, R. E. 1980. "The influence of shielding on the wind forces experienced by arrays of lattice frames." In *Proc., 5th Int. Conf. on Wind Engineering*, Fort Collins, CO, 405–420.

Willford, M. R., and A. C. Allsop. 1990. *Design guide for wind loads on unclad framed building structures during construction: Supplement 3 to the designer's guide to wind loading of building structures*. Building Research Establishment Report. Watford, UK: Building Research Establishment.

CHAPTER 6
Examples

6.1 INTRODUCTION

This chapter contains examples of how to apply the recommended guidelines for determining wind loads on the various types of structures in petrochemical and other industrial facilities.

The site for the example petrochemical facility (either planned or existing) is located in a rural area along the Texas Gulf Coast about 30 mi inland between Galveston and Corpus Christi. It is surrounded by flat open terrain that extends at least 1 mi in all directions as shown in Figure 6-1.

The velocity pressure (q_z) is determined by ASCE 7, Equation (26.10-1). Assume the following design parameters per ASCE 7:

Basic wind speed:	145 mi/h for Risk Category II, Exposure C
Exposure coefficients, K_z:	Table 26.10-1
Topographic factor, K_{zt}:	Section 26.8.2, $K_{zt} = 1.0$
Ground elevation factor, K_e:	Table 26.9-1, $K_e = 1.0$
Directionality factor, K_d:	Table 26.6-1 0.85 for main wind-force-resisting system (MWFRS), components and cladding, lattice frameworks, and trussed towers

Per ASCE 7, Section 26.6, the directionality factor shall be included in the wind loads calculated in Chapters 27 and 30.

Risk Category = II	(The typical risk category for many petrochemical and other industrial facilities is III)

Certain plant structures may be deemed as Risk Category III or IV depending on the nature of their operation (e.g., a storage tank containing highly toxic substances). The basic wind speed for Risk Category III or IV structures will be higher. See Section 26.5 of ASCE 7 for details.

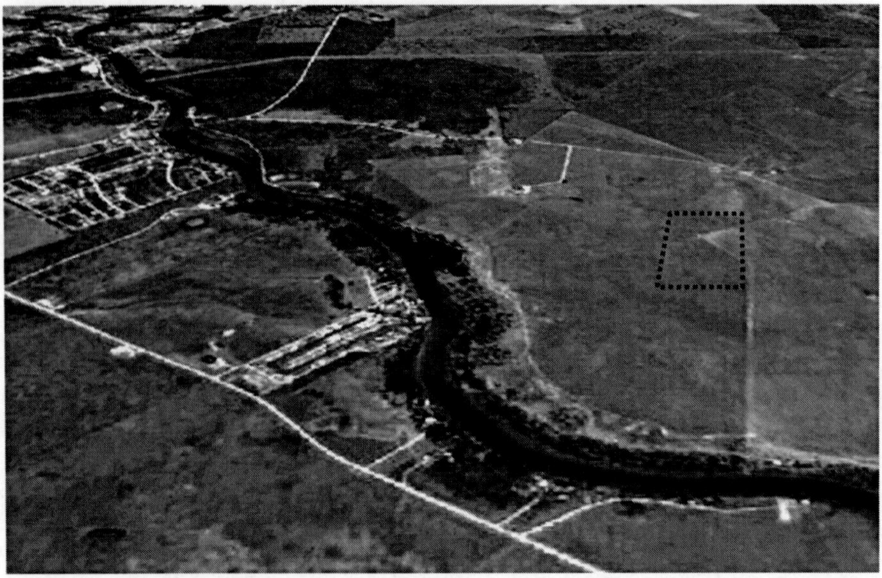

Figure 6-1. Site of petrochemical facility.
Source: USGS.

Gust effect factor, $G_f = 0.85$

Certain plant structures (e.g., a partially filled process tower) may prove to be flexible or dynamically sensitive, thus requiring a rational analysis to determine G_f. See Section 26.11 of ASCE 7 for further clarification.

6.2 PIPE RACK AND PIPE BRIDGE EXAMPLE

6.2.1 Pipe Rack

The five piping configuration cases considered are as shown in Figure 6-2 (the diameter of each pipe shown is from outside-to-outside of the insulation). Design wind forces are determined by Equation (5-1a) (repeated following), where F is the force per unit length of the piping or cable tray:

$$F = q_z G C_f A_e$$

Design wind pressure for 30 ft elevation, $K_z = 0.9823$

$$\begin{aligned} q_z &= 0.00256\ K_z\ K_{zt}\ K_d\ K_e\ V^2\ (\text{lb/ft}^2,\ V \text{ in mi/h}),\ \text{ASCE 7, Section 26.10.2} \\ &= 0.613\ K_z\ K_{zt}\ K_d\ K_e\ V^2\ (\text{N/m}^2,\ V \text{ in m/s}) \\ &= 44.94\ \text{psf}\ (2.1616\ \text{kN/m}^2) \end{aligned}$$

Gust effect factor, $G = 0.85$, ASCE 7, Section 26.11
Directionality factor $K_d = 0.85$, ASCE 7, Table 26.6-1

Force coefficients:

For structural members, $C_f = 1.8$; alternatively, for structural members above the first level, $C_f = 1.6$, and below the first level $C_f = 2.0$, Section 5.2.3
For pipes, $C_f = 0.7$, Section 5.2.3
For cable trays, $C_f = 2.0$, Section 5.2.4

Projected area:

Projected area per foot of pipe rack, A_e = Largest pipe diameter or cable tray height + 10% of the pipe or cable tray width

Sections 5.2.1 and 5.2.2

It is reasonable to limit the projected area of the pipe and cable tray to the maximum tributary height of the structure level.

$$A_e\,(\text{total}) \leq A_e(\text{max})$$

where A_e (total) is A_e (pipes) + A_e (cable tray), and A_e (max) is the maximum tributary height of that structure level.

In some cases, the pipe rack longitudinal strut or stringer might fall in the shielding envelope and can be deleted from the sum of the projected area of the piping.

6.2.1.1 *Part I: Piping and Cable Tray*

The guidelines require the consideration of the piping or cable tray separately from the structural members. The following calculations are only for piping or cable trays without the structural support members for all five configurations in Figure 6-2.

Case I: 20 ft (6.10 m) wide rack with one 48 in. (1.22 m) pipe and 14 9 in. (0.23 m) pipes

Projected area, A_e = 4 ft + (20 ft × 0.10) = 6.0 ft^2/ft (1.83 m^2/m)
Force per foot, F = 44.94 psf × 0.85 × 0.7 × 6.0 ft^2/ft = 160.4 plf (2.35 kN/m)

Case II: 20 ft (6.10 m) wide rack with 15 12 in. (0.30 m) pipes

Projected area, A_e = 1 ft + (20 ft × 0.10) = 3 ft^2/ft (0.91 m^2/m)
Force per foot, F = 44.94 psf × 0.85 × 0.7 × 3.0 ft^2/ft = 80.2 plf (1.17 kN/m)

Case III: 5 ft 6 in. (1.68 m) wide rack with one 24 in. (0.61 m) pipe and two 12 in. (0.30 m) pipes

Projected area, A_e = 2 ft + (5.5 ft × 0.10) = 2.55 ft^2/ft (0.77 m^2/m)
Force per foot, F = 44.94 psf × 0.85 × 0.7 × 2.55 ft^2/ft = 68.2 plf (0.99 kN/m)

Case IV: 4 ft 6 in. (1.37 m) wide rack with three 12 in. (0.30 m) pipes

Projected area, A_e = 1 ft + (4.5 ft × 0.10) = 1.45 ft^2/ft (0.44 m^2/m)
Force per foot, F = 44.94 psf × 0.85 × 0.7 × 1.45 ft^2/ft. = 38.8 plf (0.57 kN/ m)

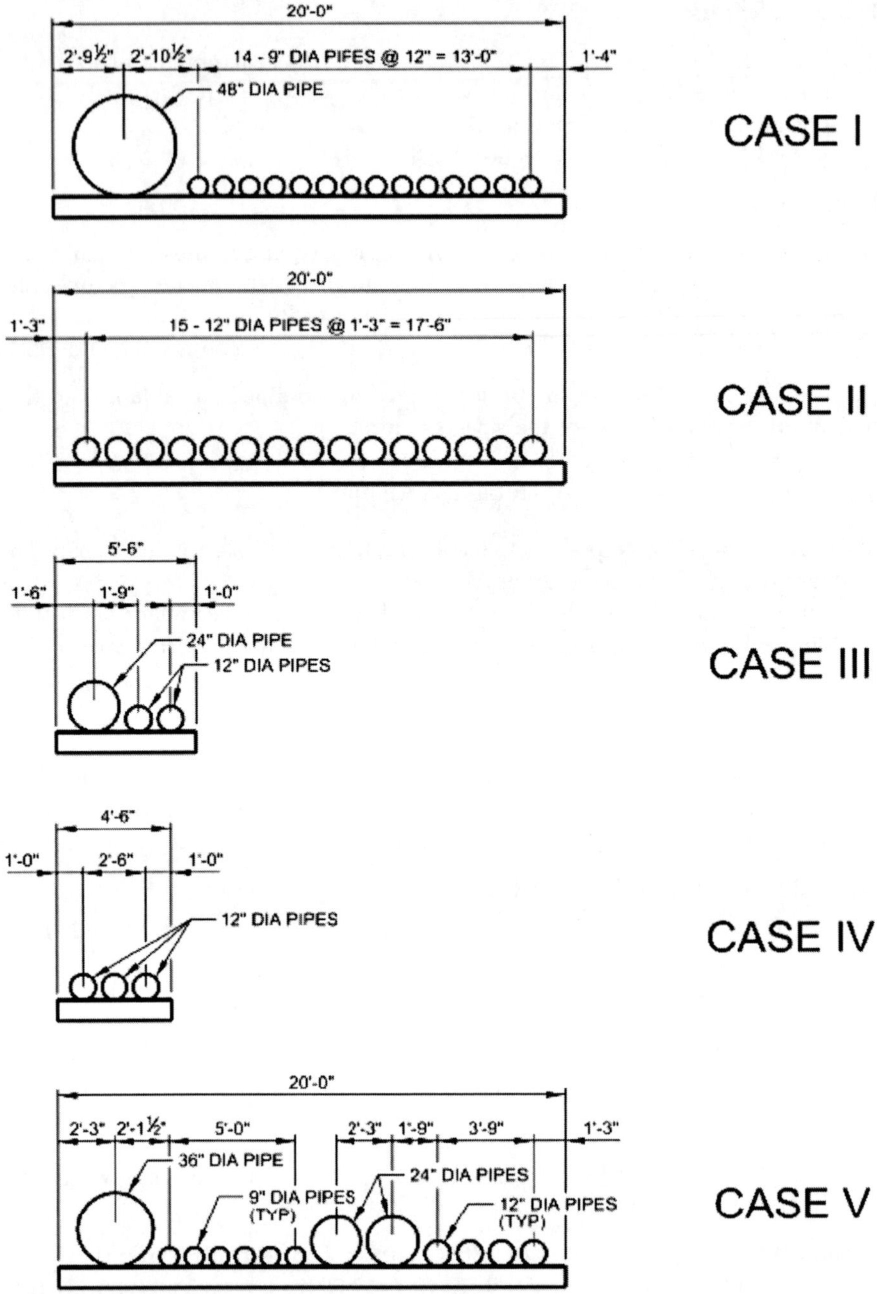

Figure 6-2. Pipe rack Example Cases I through V.

Case V: 20 ft (6.10 m) wide rack with one 36 in. (0.91 m) pipe, two 24 in. (0.61 m) pipes, four 12 in. (0.30 m) pipes, and six 9 in. (0.23 m) pipes

Projected area, A_e = 3 ft + (20 ft × 0.10) = 5.0 ft²/ft (1.52 m²/m)
Force per foot, F = 44.94 psf × 0.85 × 0.7 × 5.0 ft²/ft = 133.7 plf (1.95 kN/m)

Cable trays: 20 ft (6.10 m) wide rack with two 36 in. (0.91) trays, one 24 in. (0.61 m) tray, two 18 in. (0.46 m) trays, two 12 in. (0.30 m) trays, and two 6 in. (0.15 m) trays, all 6 in. (0.15 m) high

Projected area, A_e = 0.5 ft + (20 ft × 0.10) = 2.5 ft²/ft (0.76 m²/m)
Force per foot, F = 44.94 psf × 0.85 × 2.0 × 2.5 ft²/ft = 191.0 plf (2.79 kN/m)

6.2.1.2 Part II: Structural Members

For structural members assume the pipe rack geometry is as follows (Figure 6-3):

- 20 ft (6.10 m) wide rack with bent spacing on 20 ft (6.10 m) centers; all columns and stringers not shielded.
- The stringer and the columns are assumed to be W10 and W12 sections, respectively. Both stringer and column are assumed to have a level of fireproofing insulation as indicated in the projected area.
- Stringer member projected area per foot of rack = 1 ft²/ft (0.31 m²/m).
- Column projected area per foot of column = 1.25 ft²/ft (0.38 m²/m). (Typical concrete fireproofing thickness is 2 in., resulting in a value of 1.33 ft²/ft. This example uses a lower thickness value of fireproofing for simplicity.)
- Three levels of pipes and cable trays at elevations 18 ft (5.49 m), 24 ft (7.32 m), and 30 ft (9.15 m).
- One level of struts at 21 ft (6.40 m).

q_{18} [at elevation 18 ft (5.49 m)] = 40.4 psf (1.94 kN/m²)
q_{21} [at elevation 21 ft (6.40 m)] = 41.7 psf (2.01 kN/m²)
q_{24} [at elevation 24 ft (7.32 m)] = 42.9 psf (2.06 kN/m²)
q_{30} [at elevation 30 ft (9.15 m)] = 44.9 psf (2.16 kN/m²)

Next calculate wind loads per bent using a force coefficient C_f = 1.8 for all members per Section 5.2.3. The projected area for stringers at elevation 21 ft (6.40 m) is calculated as the sum of stringers times the stringer depth times the bent spacing.

Projected area of stringers = 2 stringers × 1 ft depth × 20 ft bent spacing
= 40 ft² (3.72 m²)

Similarly, the projected area for columns is

Projected area of columns = 2 columns × 1.25 ft width × 30 ft height
= 75 ft² (6.97 m²)

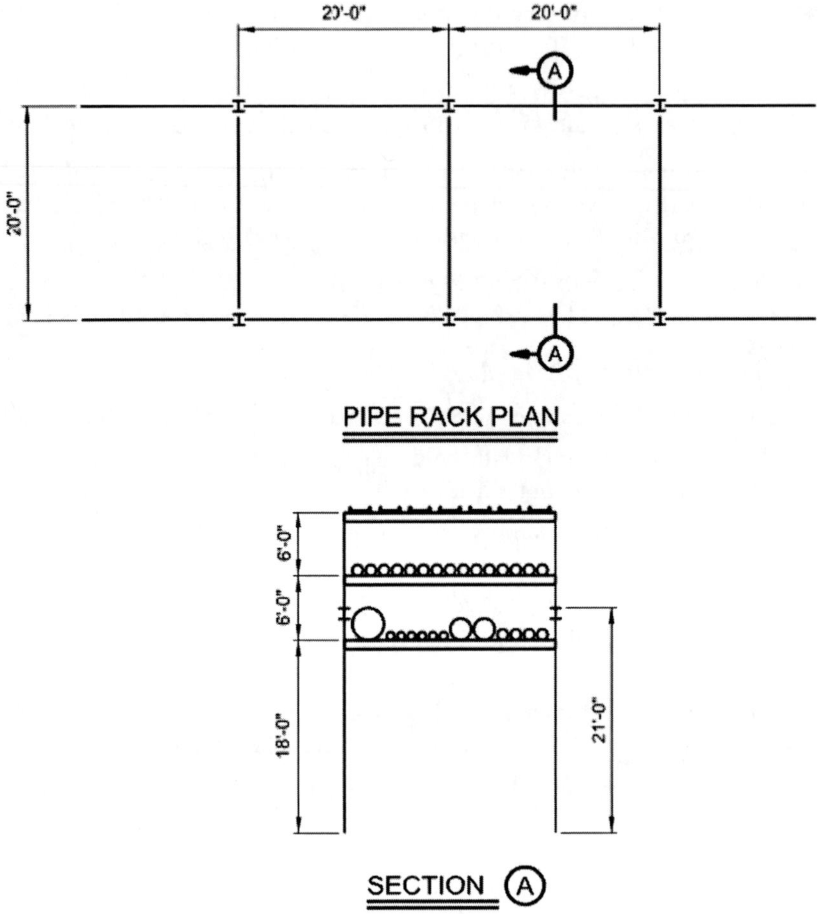

Figure 6-3. Pipe rack example framing figure.

Total force on the structural members is per Equation (5-1a)

$$F = 41.7 \text{ psf} \times 0.85 \times 1.8 \times 40 \text{ ft}^2 + 44.9 \text{ psf} \times 0.85 \times 1.8 \times 75 \text{ ft}^2$$
$$= 7{,}704 \text{ lb } (34.3 \text{ kN})$$

Alternatively, using $C_f = 1.6$ for members above the first level and $C_f = 2.0$ for members at or below the first level

Projected area for stringers $= 40 \text{ ft}^2$ (3.72 m^2)

Projected area for columns above first level $= 2$ columns $\times$ 1.25 ft width $\times$ 12 ft height $= 30 \text{ ft}^2$ (2.79 m^2)

Projected area for columns below first level $= 2$ columns $\times$ 1.25 ft width $\times$ 18 ft height $= 45 \text{ ft}^2$ (4.18 m^2)

Total force on structure F = (41.7 psf × 40 ft^2 + 44.9 psf × 30 ft^2) × 0.85 × 1.6
+ 40.4 psf × 0.85 × 45 ft^2 × 2.0
= 7,192 lb (32.0 kN)

6.2.1.3 Total Base Shear on Pipe Rack

To combine the effects of the piping, cable tray, and structural members, the pipe rack structure along with the Case V pipe configuration on the bottom level, Case II pipe configuration on the middle level, and cable trays on the top level are used (Figure 6-2).

Calculate the total base shear for the structure.

Recalculate the wind load from Part 1 for the appropriate elevation shown in Figure 6-2.

Bottom level of pipes: 133.7 plf × (40.4/44.9) × 20 ft = 2,406 lb (10.7 kN)
Middle level of pipes: 80.2 plf × (42.9/44.9) × 20 ft = 1,533 lb (6.8 kN)
Top level of cable trays: 191.0 plf × 20 ft = 3,820 lb (17 kN)
Structure = 7,704 lb (34.5 kN)
Total base shear per bent = 15,463 lb (68.8 kN)

6.2.2 Pipe Bridge Example

The pipe bridge case is similar to the pipe rack as described in Section 6.2.1. Design wind forces are determined by Equations (5-1a) (repeated below) and (5-1b), where F is the force per unit length of the piping or cable tray

Equation (5-1a): $F = q_z G C_f A_e$
Gust effect factor, G = 0.85, ASCE 7, Section 26.11
Directionality factor K_d = 0.85, ASCE, Table 26.6-1

6.2.2.1 Bridge Structural Members

For structural members assume the pipe bridge geometry is as follows (Figure 6-4):

- 20 ft (6.10 m) wide × 80 ft (24.4 m) long bridge with pipe support bents at 20 ft (6.10 m) centers.
- The top chord, bottom chord, and pipe support beams are assumed to be W10, the web columns are assumed to be W8, and the diagonals are assumed to be WT4. The pipe bridge is not fireproofed.
- Chord member projected area per foot of bridge = 0.84 ft^2/ft (0.26 m^2/m).
- Column projected area per foot of bridge = 0.67 ft^2/ft × 12 ft/10 ft = 0.80 ft^2/ft (0.24 m^2/m).
- Diagonal member projected area per foot of bridge = 0.34 ft^2/ft × 15.6 ft/10 ft = 0.53 ft^2/ft (0.16 m^2/m).
- Three levels of pipes and cable trays at elevations 24 ft (7.32 m), 30 ft (9.15 m), and 36 ft (11.0 m):

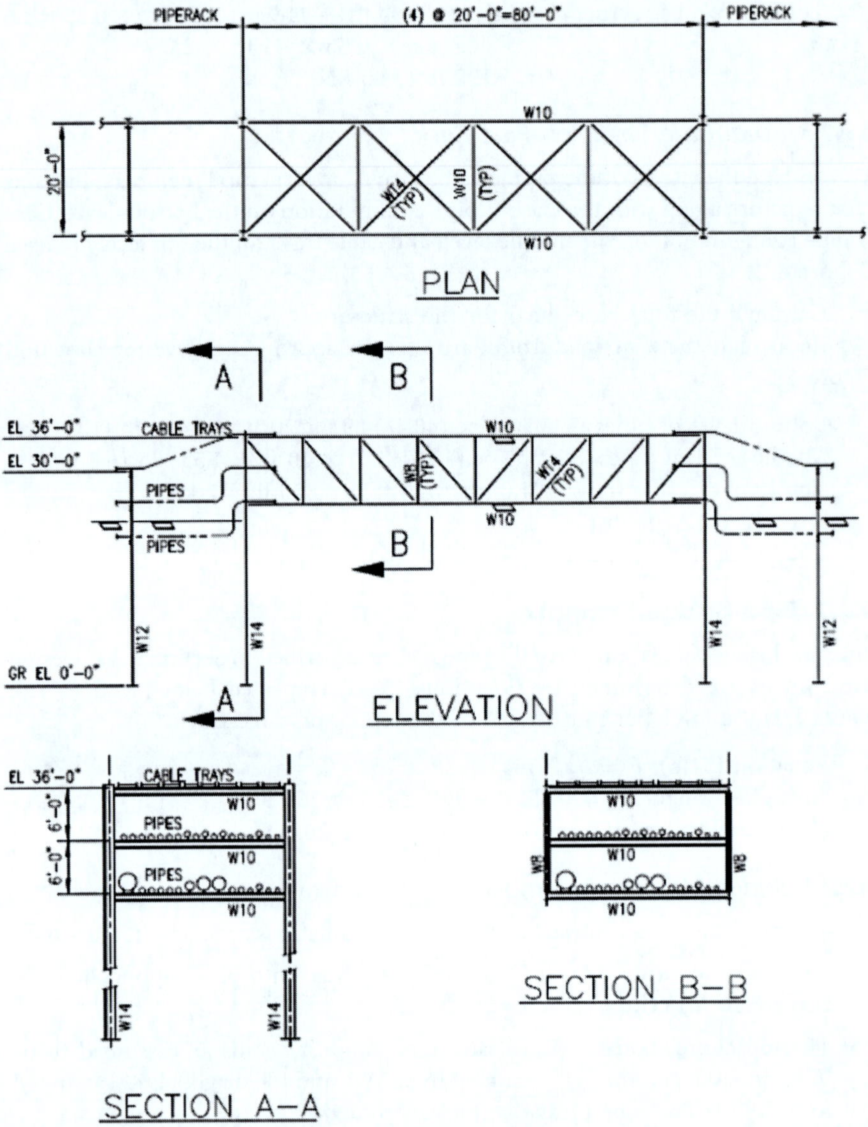

Figure 6-4. Pipe bridge example.

q_{24} [at elevation 24 ft (7.32 m)] = 42.9 psf (2.06 kN/m^2)
q_{30} [at elevation 30 ft (9.15 m)] = 44.9 psf (2.16 kN/m^2)
q_{36} [at elevation 36 ft (11.0 m)] = 46.7 psf (2.25 kN/m^2)

Next calculate wind loads on bridge members using a force coefficient $C_f = 1.6$ for all members per Section 5.2.3. In this example only members in the truss area are considered, and wind loads on column projected area below the bridge would need to be calculated for total wind loading.

6.2.2.2 *Lateral Load on Pipe Bridge*

Lateral Load on Bridge Members

All steel members are considered not to be shielded.

Projected area of half height
of bridge, $A_s = 2 \times (0.84\text{ ft/ft} + 0.8\text{ ft}^2\text{/ft/2} + 0.53\text{ ft}^2\text{/ft/2})$
$= 3.01\text{ ft}^2\text{/ft } (0.91\text{ m}^2\text{/m})$

Force on top chord, $F = 46.7\text{ psf} \times 0.85 \times 1.6 \times 3.0\text{ ft}^2\text{/ft} \times 80\text{ ft}$
$= 15{,}294\text{ lb } (68\text{ kN})$

Force on bottom chord, $F = 42.9\text{ psf} \times 0.85 \times 1.6 \times 3.0\text{ ft}^2\text{/ft} \times 80\text{ ft}$
$= 14{,}049\text{ lb } (62.5\text{ kN})$

Lateral loads on bridge members, $F = 15{,}294\text{ lb} + 14{,}049\text{ lb}$
$= 29{,}343\text{ lb } (130.5\text{ kN})$

Alternatively, apply Section 5.3 for open-frame structures to generate lateral loads on the bridge. In this case shielding is considered.

For bridge lateral surface
$A_g = 12\text{ ft}^2\text{/ft} + 0.84\text{ ft}^2\text{/ft} = 12.84\text{ ft}^2\text{/ft } (3.91\text{ m}^2\text{/m})$

Effective solid area: $A_s = 0.84\text{ ft}^2\text{/ft} \times 2 + 0.80\text{ ft}^2\text{/ft} + 0.53\text{ ft}^2\text{/ft}$
$= 3.01\text{ ft}^2\text{/ft } (0.92\text{ m}^2\text{/m})$
$= A_e$

Solidity ratio: $\varepsilon = A_s/A_g$
$= 3.01\text{ ft}^2\text{/ft}/12.84\text{ ft}^2\text{/ft} = 0.234$
$N = 2$
$S_f = 20\text{ ft } (6.1\text{ m})$
$B = 80\text{ ft } (24.4\text{ m})$
$S_f/B = 20/80 = 0.25$

From Figure 5-1, $C_{Dg} = 0.60$
From Equation (5-2), $C_f = C_{Dg}/\varepsilon = 0.60/0.234 = 2.56$

Lateral loads on bridge members

$$F = q_z\, G\, C_f\, A_e \tag{5-1a}$$
$= (46.7\text{ psf} + 42.9\text{ psf})/2 \times 0.85 \times 2.56 \times 3.01\text{ ft}^2\text{/ft} \times 80\text{ ft}$
$= 23{,}469\text{ lb } (104.4\text{ kN})$

Lateral Load on Pipes and Cable Trays

The pipe support beams in the bridge are similar to those in the pipe rack example. Case V pipes on the bottom level, Case II pipes on the middle level, and cable trays on the top level are used (Figure 6-3).

Bottom level of pipes: $F = 133.7\text{ plf} \times (42.9/44.9) \times 80\text{ ft} = 10{,}220\text{ lb } (45.5\text{ kN})$
Middle level of pipes: $F = 80.2\text{ plf} \times (44.9/44.9) \times 80\text{ ft} = 6{,}416\text{ lb } (28.5\text{ kN})$
Top level of cable trays: $F = 191\text{ plf } (46.7/44.9) \times 80\text{ ft} = 15{,}893\text{ lb } (70.7\text{ kN})$

Total Lateral Load on Pipe Bridge

Using the first of the two methods for calculating the force on the structure

$$F = 29{,}343 \text{ lb} + 10{,}220 \text{ lb} + 6{,}416 \text{ lb} + 15{,}893 \text{ lb}$$
$$= 61{,}871 \text{ lb } (275.2 \text{ kN})$$

Alternatively, the second method for calculating the force on the structure could be used.

6.2.2.3 Longitudinal Load on Pipe Bridge

Longitudinal Load on Bridge Members

Apply Section 5.3 for open-frame structures to generate longitudinal loads for bridge members using an interior truss bay to define the projected area.

Bridge cross section: $A_g = 20.83 \text{ ft} \times 12.83 \text{ ft} = 267.3 \text{ ft}^2 \ (24.8 \text{ m}^2)$

Solid area of bridge frame: $A_{s1} = 0.83 \text{ ft} \times 20 \text{ ft} \times 3 + 0.67 \text{ ft} \times 12 \text{ ft} \times 2$
$= 65.9 \text{ ft}^2 \ (6.1 \text{ m}^2)$

Solid projected area of diagonal bracing: $A_{s2} = 0.33 \text{ ft} \times 12 \text{ ft} \times 2$
$= 7.92 \text{ ft}^2 \ (0.74 \text{ m}^2)$

Effective solid area: $A_s = A_{s1} + A_{s2}$
$= 65.9 \text{ ft}^2 + 7.92 \text{ ft}^2 = 73.8 \text{ ft}^2 \ (6.86 \text{ m}^2)$
$= A_e$

Solidity ratio: $\varepsilon = A_s/A_g$
$= 73.8 \text{ ft}^2/267.3 \text{ ft}^2 = 0.276$
$N = 9$
$S_f = 10 \text{ ft } (3.1 \text{ m})$
$B = 20.83 \text{ ft } (6.4 \text{ m})$
$S_f/B = 10/20.83 = 0.48$

From Figure 5-1, $C_{Dg} = 2.18$
From Equation (5-2), $C_f = C_{Dg}/\varepsilon = 2.18/0.276 = 7.89$

The longitudinal loads on bridge members Equation (5-1a):

$$F = q_z \, G \, C_f \, A_e$$
$$= (46.7 \text{ psf} + 42.9 \text{ psf})/2 \times 0.85 \times 7.89 \times 73.8 \text{ ft}^2$$
$$= 22{,}186 \text{ lb } (98.7 \text{ kN})$$

The shielding of pipes at the end of the pipe bridge will reduce the longitudinal wind load on the bridge structure. At present, however, proven laboratory testing to determine the shielding results is lacking. Therefore, this shielding effect has not been addressed in this example.

Longitudinal Load on Pipes and Cable Trays

Assume all piping is turned up from the pipe rack to the pipe bridge at the beginning of the bridge and returns down to the pipe rack at the end of the bridge.

Also assume half of the pipe and cable tray loads are applied to the pipe bridge and the other half to the pipe rack.

Bottom level of pipes:

Solid area of pipes = (36 in./12 in./ft + 2 × 24 in./12 in./ft + 4 × 12 in./12in./ft + 6 × 9 in./12 in./ft) × 6 ft = 93.0 ft^2 (8.6 m^2)

The center-to-center distance of pipes is less than three diameters; therefore, C_f should be increased 20% per Section 5.5.1.4

F = 42.9 psf × 0.85 × 0.7 × 1.2 × 93.0 ft^2 × 2 = 5,697 lb (25.3 kN)

Middle level of pipes:

Solid area of pipes = (15 × 12 in./12 in./ft) × 6 ft = 90.0 ft^2 (8.4 m^2)
F = 44.9 psf × 0.85 × 0.7 × 1.2 × 90.0 ft^2 × 2 = 5,771 lb (25.7 kN)

Top level of cable trays: Assume cable trays and cables occupied 90% of the cross-sectional area of the pipe rack and C_f = 1.0 in this case

Solid area of cable trays = 0.9 × 20 ft × 6 ft = 108.0 ft^2 (10.0 m^2)
F = 46.7 psf × 0.85 × 1.0 × 108 ft^2 × 2 ft = 8,574 lb (38.1 kN)

Total Longitudinal Load on Pipe Bridge

F = 22,186 lb + 5,697 lb/2 + 5,771 lb/2 + 8,574 lb/2
= 32,207 lb (143.3 kN)

6.3 OPEN-FRAME EXAMPLES

In this section, the wind load will be determined on an open-frame structure using the three methods described in Section 5.3, Appendix 5A, and Appendix 5B.

The structure considered is shown in Figures 6-5, 6-6, and 6-7. Design wind forces are determined by Equation (5-la)

$$F = q_z G C_f A_e$$

The velocity pressure q_z is determined using Section 26.10.2 of ASCE 7

$$q_z = 0.00256 K_z K_{zt} K_d K_e V^2 \qquad (26.10\text{-}1)$$

Refer to Section 6.1 for the values to be used for the example. It is convenient to determine the velocity pressures at the midfloor heights and at the top of the structure. Table 6-1 summarizes the determined values for K_z and q_z.

The gust effect factor is determined next. The ratio of height to least horizontal dimension = 83 ft/41 ft = 2.02 < 4; therefore, the structure is not considered a flexible structure. Use G = 0.85, as described in Section 6.1.

Table 6-1. Values for q_z and K_z.

Height aboveground z (ft)	K_z	q_z *(psf)*
10	0.85	38.9
34	1.01	46.1
65	1.16	52.9
h = 83	1.22	55.7

Note: To convert psf to N/m^2, multiply values in this table by 47.88. Although the top of the third-floor level is at 82 ft, the structure height *h* was increased slightly to account for the handrail and minor equipment on top of the structure (Figures 6-5 and 6-6).

6.3.1 Open-Frame Example Using Method of Section 5.3

6.3.1.1 Along-Wind Force Calculations (Wind Toward Frame 3)

To calculate the force coefficient, the solidity ratio ε must first be computed from Equation (5-3), which stated $\varepsilon = A_S/A_g$. The gross area (or envelope area) is the area within the outmost projections of the front face normal to the wind direction. The width used here is measured from outside column face to outside column face. For the wind direction shown in Figure 6-5

$$A_g = 83 \text{ ft (height)} \times 41 \text{ ft (width)} = 3{,}403 \text{ ft}^2 \ (316 \text{ m}^2)$$

To determine the effective solid area, the solid area of the windward frame must first be calculated per Section 5.3.4.1. To facilitate the computation of forces later in the problem, it is convenient to calculate the solid areas from midfloor to midfloor and sum these to obtain the total solid area of the frame. Member sizes are given in the respective figures (except details of handrails and stairs). Calculation of the solid area of the windward frame (Column Line 3) is summarized in Table 6-2. The stairs were considered part of the windward frame (Figure 6-5). The stairs column in Table 6-2 includes areas of stair stringers, struts, handrails, and bracing.

Because the middle and leeward frames (Column Lines 2 and 1, respectively) are similar to the windward frame with the exception of not having stairs, the solid areas and hence the solidity ratios for these two frames will be less than the windward frame, so A_S is equal to the solid area of windward frame per Section 5.3.4.4, which leads to

$$\varepsilon = A_S/A_g = 1{,}378 \text{ ft}_2/3{,}403 \text{ ft}_2 = 0.405$$

Next, the coefficient C_{Dg} is obtained from curves given in Figure 5-1 as a function of the solidity ratio ε, the number of frames *N*, and the frame spacing

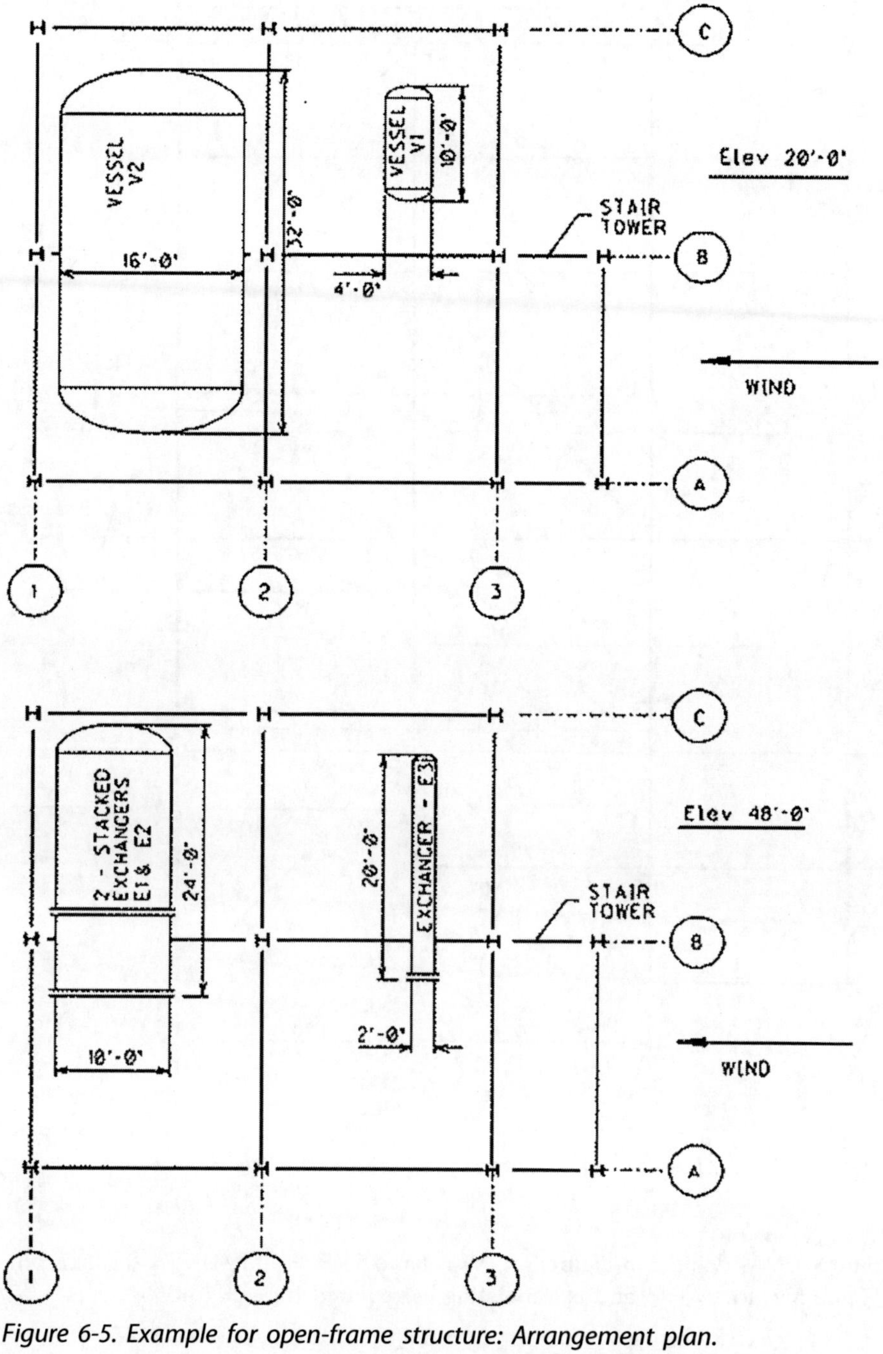

Figure 6-5. Example for open-frame structure: Arrangement plan.

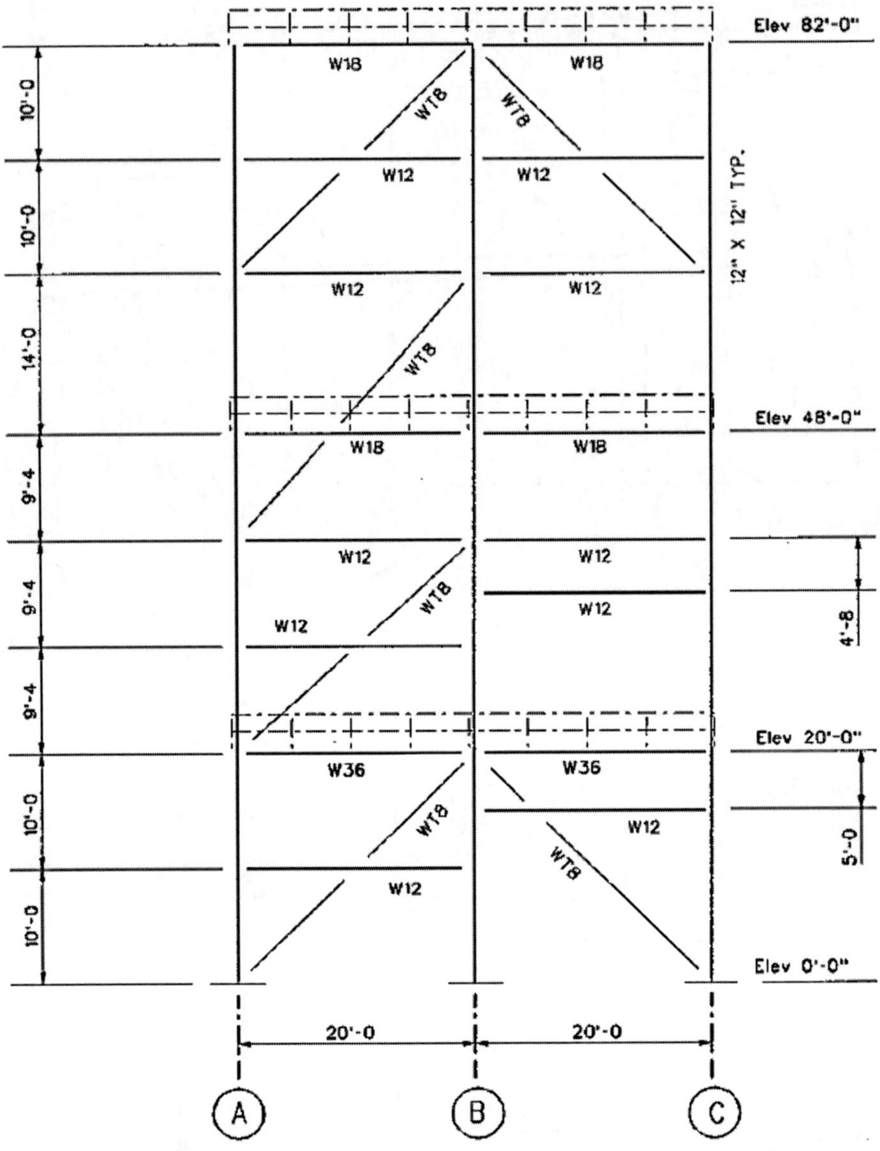

Figure 6-6. Example for open-frame structure: Column Line 3.

ratio S_F/B. As defined in Figure 5-1, $N = 3$ and $S_F/B = 20$ ft/41 ft $= 0.488$. From Figure 5-1, for $N = 3$ and extrapolating as required for $\varepsilon = 0.405$

$C_{Dg} = 1.18$, for $S_F/B = 0.5$ for $N = 3$
$C_{Dg} = 1.12$, for $S_F/B = 0.33$ for $N = 3$

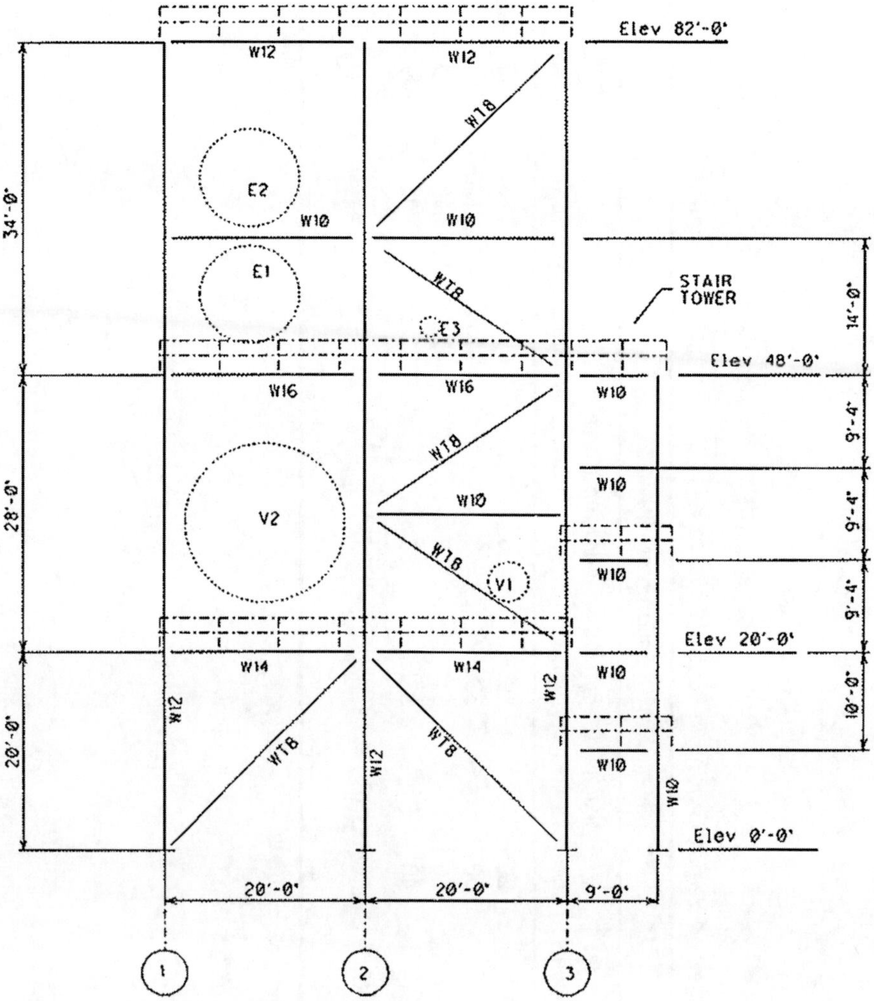

Figure 6-7. Example for open-frame structure: Column Line A.

Interpolating for $S_F/B = 0.488$

$$C_{Dg} = 1.12 + (1.18 - 1.12)\ [(0.488 - 0.33)/(0.5 - 0.33)] = 1.176$$

Next, the gross area force coefficient C_{Dg} is converted to a force coefficient compatible with ASCE 7 by means of Equation (5-2)

$$C_f = C_{Dg}/\varepsilon = 1.176/0.405 = 2.90$$

The force coefficient could also be determined directly using the alternate method of Appendix 5A. The alternate method is somewhat more time consuming, but it covers a wider range of solidity ratios and frame spacing ratios and can also be used to determine more accurate load combinations (See Section 6.3.2).

Table 6-2. Solid Area of Windward Frame and Vertical Bracing in Bents Parallel to Wind Direction (Wind toward Frame 3).

		Solid areas (ft^2)							
Floor level	*Tributary height (ft)*	*Columns*	*Beams*	*Intermediate beams*	*Bracing (normal)*	*Bracing (three parallel bents)*	*Handrails*	*Stairs*	*Total*
0	0–10 = 10	30	0	20	19	20	0	76	165
1	10–34 = 24	72	120	60	33	48	32	150	515
2	34–65 = 31	93	60	80	31	62	32	91	449
3	65–83 = 18	51	60	40	32	34	32	0	249
Total solid area of windward frame and parallel bents (ft^2) =									1,378 ($128\ m^2$)

Note: To convert ft^2 to m^2, multiply values in this table by 0.0929.

Table 6-3. Total Force, Structural Frame and Appurtenances, F_S *(Wind toward Frame 3).*

Floor level	q_z *(psf)*	G	C_f	A_e *(ft²)*	η_{floor}	F *(lb)*
0	38.9	0.85	2.90	165	–	15,822
1	46.1	0.85	2.90	515	0.953	55,772
2	52.9	0.85	2.90	449	0.973	56,968
3	55.7	0.85	2.90	249	1.0	34,176
	$F_S = \sum F =$					162,737

Note: To convert pound force (lbf) to N, multiply *F* values in this table by 4.448.

The area of application of force A_e has already been determined per floor level during calculation of the solidity ratio. The wind force transmitted to each floor level may now be found by Equation (5-1a), $F = q_z G C_f A_e$, as shown in the following. For floor levels that have solid flooring, the wind load on these levels may be reduced by the shielding factor η_{floor}, per Section 5.3.4.2. For this example, assume that the lower two levels are decked with checkered plate (solid flooring) and that the top level is floored with grating. Applying Equation (5-4), $\eta_{\text{floor}} = 1 - 0.2(A_{fb}/A_S)$

Level 1: $A_{fb} = 120\ \text{ft}^2$ $A_S = 515\ \text{ft}^2$, $\eta_{\text{floor}} = 0.953$
Level 2: $A_{fb} = 60\ \text{ft}^2$ $A_S = 449\ \text{ft}^2$, $\eta_{\text{floor}} = 0.973$
Level 3: $\eta_{\text{floor}} = 1.0$

The total force on the structural frame and appurtenances, F_S, is 162.7 kips (723.9 kN), found by summing the forces at all levels in Table 6-3.

These forces are because of wind acting on the frame only. Wind forces acting on the vessels, equipment, and piping are computed in accordance with Sections 5.2 and 5.5 and added at the levels where the items are located. The structure supports two horizontal vessels on Floor 1 and three horizontal exchangers on Floor 2 (see respective figures). All these items are subject to transverse wind loads for the wind direction under consideration. The projected diameter is equal to the vessel insulated diameter plus 1.5 ft (0.46 m) per Section 5.5.2.2. The force coefficient is determined from ASCE 7, Figure 29.4-1. The vessels and exchangers were assumed to be moderately smooth. These properties are listed in Table 6-4.

The wind load on a given vessel or exchanger is the sum of the wind load on the supports, platforms, large connecting pipes, and the cylinder itself as summarized in Table 6-5. This example only considers the load on the cylinder. For determining q_h, the height of each vessel was assumed to be the midfloor elevation above the supporting floor level. A small improvement in accuracy could be obtained by using the actual top elevation for each piece of equipment.

The projected area of piping and electrical was assumed as 20% of the projected vessel area. For example, on Floor 1, the piping area is equal to 0.2 (55 + 560) = 123 ft² (11.42 m²). The piping was assumed to have a moderately

Table 6-4. Equipment Properties (Wind toward Frame 3).

Equipment	*Floor level*	*Vessel diameter (ft)*	*Projected diameter* D *(ft)*	*Length* B *(ft)*	*Projected area* A_e *(ft²)*	B/D	C_f
V1	1	4	5.5	10	55	1.8	0.51
V2	1	16	17.5	32	560	1.8	0.51
E1	2	10	11.5	24	276	2.1	0.52
E2	2	10	11.5	24	276	2.1	0.52
E3	2	2	3.5	20	70	5.7	0.58

Note: To convert ft^2 to m^2, multiply A_e values in this table by 0.0929.

Table 6-5. Gross Wind Forces: Equipment (Wind toward Frame 3).

Equipment	q_h *(psf)*	G	C_f	A_e *(ft²)*	F *(lb)*
V1	46.1	0.85	0.51	55	1,099
V2	46.1	0.85	0.51	560	11,191
E1	52.9	0.85	0.52	276	6,453
E2	52.9	0.85	0.52	276	6,453
E3	52.9	0.85	0.58	70	1,826

Note: To convert pound force (lb) to N, multiply *F* values in this table by 4.448.

smooth surface. The force coefficient is determined from ASCE 7, Figure 29.4-1 (also see Section 5.2.3). The wind load on piping per floor level is summarized in Table 6-6.

The total equipment load per floor level is equal to the sum of all of the vessel, exchanger, and piping wind loads on that floor level. For purposes of determining the overall wind load on the structure, equipment and piping loads can be reduced to account for shielding effects (shielding of the equipment by upwind portions of the structure, shielding of portions of the structure by upwind equipment, and equipment-to-equipment shielding). For purposes of designing individual vessels and supports, the loads should not be reduced.

Because the vessels and exchangers are in the wind shadow of the stairs, vertical bracing, middle column, and intermediate beams, it is appropriate to reduce the equipment load for shielding by the upwind frames. Per Equation (5-5), the shielding factor η_{equip} is given by

$$\eta_{equip} = \exp[-1.4(C_f \cdot \varepsilon)^{1.5}]$$

where $\varepsilon = 0.405$ and $C_f = 2.90$, as calculated previously

Table 6-6. Gross Wind Forces: Piping Per Level (Wind toward Frame 3).

Floor level	q_h *(psf)*	G	C_f	A_e *(ft²)*	F *(lb)*
1	46.1	0.85	0.7	123	3,374
2	52.9	0.85	0.7	124	3,903

Note: To convert pound force (lb) to N, multiply *F* values in this table by 4.448.

Table 6-7. Gross Wind Force: Equipment and Piping (Wind Toward Frame 3).

Floor level	*Equipment and piping load*	η_{equip}	*Reduced load [lb (kN)]*
1	V1 + V2 + piping = 1,099 + 11,191 + 3,374 = 15,664 lb (69.7 kN)	0.17	2,663 (11.8)
2	E1 + E2 + E3 + piping = 6,453 + 6,453 + 1,826 + 3,903 = 18,635 lb (82.9 kN)	0.17	3,168 (14.1)
	Total equipment load = $\sum$ Reduced loads =		5,831 (25.9)

$$\eta_{equip} = \exp[-1.4(2.90 \times 0.405)^{1.5}] = 0.17$$

If the solidity of the upwind frames varies considerably from level to level, it would be appropriate to calculate ε and C_f for each level supporting equipment rather than using a single value of ε and C_f for the overall structure.

Summing the loads on the vessels, exchangers, and piping per level and applying the shielding factor yields the total wind load on equipment in Table 6-7.

6.3.1.2 Crosswind Force Calculations (Wind Toward Frame A)

The next step is to repeat the analysis for the wind direction normal to Column Line A (Figures 6-5 and 6-7). The member sizes are the same on this elevation except that the intermediate beams are W10s and

Beams EL 20 ft 0 in., W14
Beams EL 48 ft 0 in., W16
Beams EL 82 ft 0 in., W12

The gross area of the windward face includes the stair tower on the right side of the structure

$$A_g = (83 \times 41) + (9 \times 49) = 3{,}844 \text{ ft}^2$$

The solid areas for the windward frame are given in Table 6-8. The stairs column in the table includes the areas of the stair column, struts, and handrails.

Table 6-8. Solid Area of Windward Frame and Vertical Bracing in Bents Parallel to Wind Direction (Wind Toward Frame A).

		Solid areas (ft^2)							
Floor level	*Tributary height (ft)*	*Columns*	*Beams*	*Intermediate beams*	*Bracing (normal)*	*Bracing (three parallel bents)*	*Handrails*	*Stairs*	*Total*
0	0–10 = 10	30	0	0	19	20	0	24	93
1	10–34 = 24	72	46	8	35	48	32	44	285
2	34–65 = 31	93	53	41	36	62	32	36	353
3	65–83 = 18	51	40	0	16	34	32	0	173
Total solid area of windward frame and parallel bents (ft^2) =									904 ($84\ m^2$)

Note: To convert ft^2 to m^2, multiply values in this table by 0.0929.

Because the solidity of neither the middle nor leeward frames (Column Lines B and C, respectively) exceeds that of the windward frame, A_S is equal to the solid area of windward frame, yielding

$$\varepsilon = A_S/A_g = 904 \text{ ft}^2/3{,}844 \text{ ft}^2 = 0.235$$

The frame spacing ratio in this direction is $S_F/B = 20$ ft/46 ft $= 0.435$. Because the width is not uniform (the stair tower stops at the second-floor level), an average value of B was used. From Figure 5-1 for $N = 3$ and $\varepsilon = 0.235$

$$C_{Dg} = 0.82 \text{ for } S_F/B = 0.5$$
$$C_{Dg} = 0.79 \text{ for } S_F/B = 0.33$$

Therefore, use $C_{Dg} = 0.809$

$$C_f = C_{Dg}/\epsilon = 0.809/0.235 = 3.44$$

As was done for the orthogonal direction, the shielding factor, η_{floor}, is determined for each floor level by applying Equation (5-4), $\eta_{floor} = 1 - 0.2(A_{fb}/A_s)$.

Level 1: $A_{fb} = 46$ ft^2 $A_S = 285$ ft^2, $\eta_{floor} = 0.968$
Level 2: $A_{fb} = 53$ ft^2 $A_S = 353$ ft^2, $\eta_{floor} = 0.970$
Level 3: $\eta_{floor} = 1.0$

The wind forces per floor level are given in Table 6-9.

The wind direction is parallel to the axis of the vessels and exchangers (longitudinal wind). The vessels have rounded heads and the exchangers have flat heads. Force coefficients for this case are given in Section 5.5.2.4, and the wind loads are listed in Table 6-10.

The wind load on the piping and electrical is the same as calculated previously. For this wind direction, there is no significant shielding of the equipment by the windward frame and no equipment-to-equipment shielding, so no reduction is taken on equipment load. Summing the loads on the vessels, exchangers, and piping per level yields

Level 1: V1 + V2 + piping = 470 + 4,722 + 3,374 = 8,566 lb (38.1 kN)
Level 2: E1 + E2 + E3 + piping = 5,612 + 5,612 + 540 + 3,903 = 15,666 lb (69.7 kN)

for a total equipment and piping load of $F_E = 24{,}232$ lb (107.8 kN).

Table 6-9. Total Force, Structural Frame and Appurtenances, F_S (Wind Toward Frame A).

Floor level	q_z *(psf)*	G	C_f	A_e *(ft^2)*	η_{floor}	F *(lb)*
0	38.9	0.85	3.44	93	—	10,578
1	46.1	0.85	3.44	285	0.968	37,188
2	52.9	0.85	3.44	353	0.970	52,964
3	55.7	0.85	3.44	173	1.0	28,176
	$F_S = \sum F =$					128,906

Note: To convert pound force (lb) to N, multiply *F* values in this table by 4.448.

6.3.1.3 Summary

The results thus far are summarized in Table 6-11. The load combinations for design are the application of F_T in one direction simultaneously with $0.5F_S$ in the other, per Section 5.3.6.1. These combinations are shown in Figure 6-8.

Table 6-10. Gross Wind Forces, Equipment (Wind Toward Frame A).

Equipment	q_h *(psf)*	G	C_f	A_e *(ft²)*	F *(lb)*
V1	46.1	0.85	0.5	24	470
V2	46.1	0.85	0.5	241	4,722
E1	52.9	0.85	1.2	104	5,612
E2	52.9	0.85	1.2	104	5,612
E3	52.9	0.85	1.2	10	540

Note: To convert pound force (lb) to N, multiply *F* values in this table by 4.448.

Table 6-11. Summary.

Wind loading	*Wind, Direction 1 [kips (kN)]*	*Wind, Direction 2 [kips (kN)]*
Wind load on structural frame F_S	162.7 (724)	128.9 (573)
Wind load on equipment and piping F_E	5.8 (26)	24.2 (108)
Total wind load on structure F_T	168.5 (750)	153.1 (681)

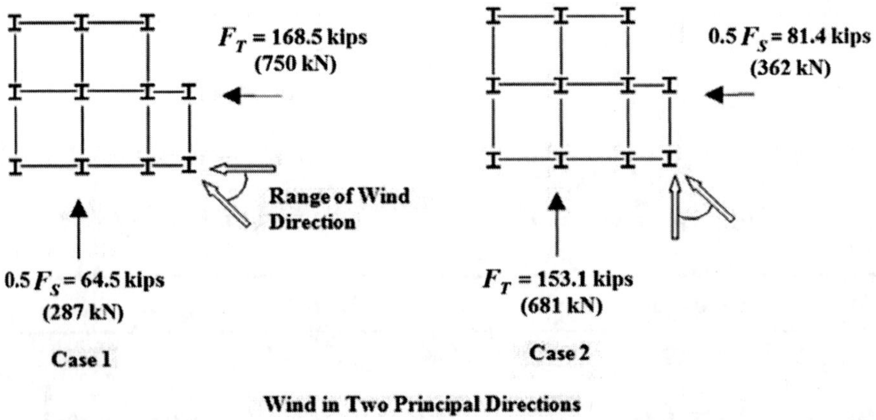

Figure 6-8. Design load cases for open-frame example (using method of Section 5.3).

This example only determines the *total* wind load in each direction. For actual design of the structure, the wind load calculation must be refined. The wind load on the structure frame (F_S) in each direction would need to be determined for each structure node by considering A_e for each individual node. The wind load on equipment and piping (F_E) would need to be applied as appropriate to each individual item of equipment. This refinement is beyond the scope of this example.

6.3.2 Open-Frame Example Using Method of Appendix 5A

The example problem of Section 6.3.1 will be reworked using the method of Appendix 5A. Refer to Section 6.3.1 as necessary for the previously calculated values.

6.3.2.1 Wind toward Frame 3

$\varepsilon = 0.405$ $\quad S_F/B = 0.488$ $\quad N = 3$

From Equation (5A-1), $\alpha_{max} = (10 + 58\varepsilon)° = 33.5°$

From Figure 5A-3, for $\alpha_{max} = 33.5°$, $N = 3$, and $S_F/B = 0.488$

For $\varepsilon = 0.1$, $C_f = 3.52$
For $\varepsilon = 0.5$, $C_f = 2.53$

Interpolating for $\varepsilon = 0.405$

$$C_f = 3.52 - [(0.405 - 0.1) / (0.5 - 0.1) \times (3.52 - 2.53)] = 2.77$$

Referring to Table 6-3: $F_s = (2.77 / 2.90) \times 162.7$ kips $= 155.4$ kips
Referring to Table 6-7: $F_E = 5.8$ kips

$\therefore F_T = 155.4$ kips $+ 5.8$ kips $= 161.2$ kips

Considering perpendicular direction

$$90° - \alpha_{max} = 90° - 33.5° = 56.5°$$

$\varepsilon = 0.235$ $\quad S_F/B = 0.435$ $\quad N = 3$

From Figure 5A-3, for $\alpha = 56.5°$, $N = 3$, and $S_F/B = 0.435$
For $\varepsilon = 0.1$, $C_f = 2.25$
For $\varepsilon = 0.5$, $C_f = 1.76$

Interpolating for $\varepsilon = 0.235$

$$C_f = 2.25 - [(0.235 - 0.1) / (0.5 - 0.1) \times (2.25 - 1.76)] = 2.08$$

Referring to Table 6-9: $F_S = (2.08 / 3.44) \times 128.9$ kips $= 77.9$ kips

6.3.2.2 Wind Toward Frame A

$\varepsilon = 0.235$ $\quad S_F/B = 0.435$ $\quad N = 3$

From Equation (5A-1), $\alpha_{max} = (10 + 58\varepsilon)° = 23.6°$
From Figure 5A-3, for $\alpha_{max} = 23.6°$, $N = 3$, and $S_F/B = 0.435$

For $\varepsilon = 0.1$, $C_f = 3.80$
For $\epsilon = 0.5$, $C_f = 2.34$

Interpolating for $\varepsilon = 0.235$

$$C_f = 3.80 - [(0.235 - 0.1) / (0.5 - 0.1) \times (3.80 - 2.34)] = 3.31$$

Referring to Table 6-9: $F_s = (3.31 / 3.44) \times 128.9$ kips $= 124.0$ kips
Referring to Tables 6-6 and 6-10 and Section 6.3.1.2: $F_E = 24.2$ kips

$$\therefore F_T = 124.0 \text{ kips} + 24.2 \text{ kips} = 148.2 \text{ kips}$$

Considering perpendicular direction

$$90° - \alpha_{max} = 90° - 23.6° = 66.4°$$

$\varepsilon = 0.405$ $\quad S_F/B = 0.488$ $\quad N = 3$

From Figure 5A-3, for $\alpha = 66.4°$, $N = 3$, and $S_F/B = 0.488$
For $\varepsilon = 0.1$, $C_f = 1.42$
For $\varepsilon = 0.5$, $C_f = 1.27$

Interpolating for $\varepsilon = 0.405$

$$C_f = 1.42 - [(0.405 - 0.1) / (0.5 - 0.1) \times (1.42 - 1.27)] = 1.31$$

Referring to Table 6-3: $F_S = (1.31 / 2.90) \times 162.7$ kips $= 73.5$ kips

6.3.2.3 Summary

The results of this example problem are shown in Figure 6-9. Calculation of loads on equipment is identical to the method used in the previous example.

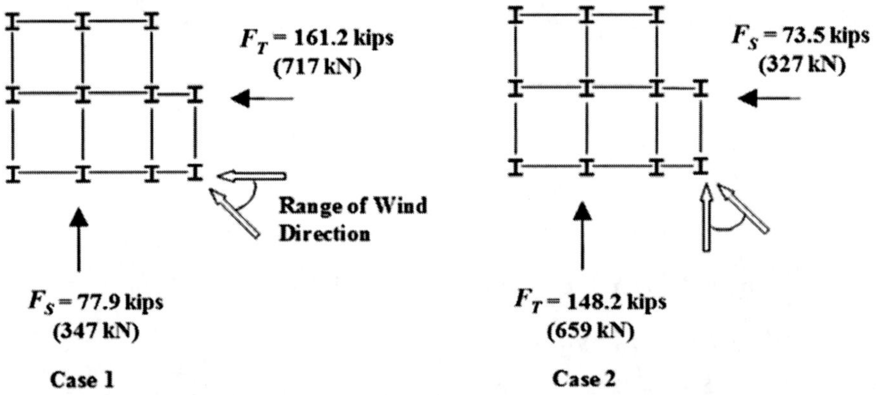

Figure 6-9. Design load cases for open-frame example (using method of Appendix 5A).

6.3.3 Open-Frame Example Using Method of Appendix 5B

The example problem of Section 6.3 will be reworked using the method of Appendix 5B, assuming the structure to be a high-solidity open-frame structure. Refer to Section 6.3.1 as necessary for the previously calculated values.

6.3.3.1 Wind Toward Frame 3

$$L = 1/2 \times [(40 \text{ ft} + 1 \text{ ft}) + (49 \text{ ft} + 1 \text{ ft})] = 45.5 \text{ ft} \approx 46 \text{ ft (average)}$$
$$B = 41 \text{ ft}$$
$$A_g = 3403 \text{ ft}^2$$

$$L/B = 46/41 = 1.12 < 1.5$$

$$\therefore C_f = 1/4 \times (L/B)^2 + 1.4$$
$$= 1/4 \times (1.12)^2 + 1.4 = 1.71$$

The total wind force on the structure, F_T, will be calculated by application of Equation (5-1a), as shown in Table 6-12. The calculated force, F_T, includes the wind load on the structure and appurtenances, equipment, piping, and others.

6.3.3.2 Wind Toward Frame A

$$L = 41 \text{ ft}$$
$$B = 46 \text{ ft}$$
$$A_g = 3{,}844 \text{ ft}^2$$

$$L/B = 41/46 = 0.89 < 1.5$$

$$\therefore C_f = 1/4 \times (L/B)^2 + 1.4$$
$$= 1/4 \times (0.89)^2 + 1.4 = 1.60$$

The total wind force on the structure, F_T, in this direction will be calculated similarly to that in Section 6.3.3.1 and is given in Table 6-13.

6.3.3.3 Summary

The results of this example problem are shown in Figure 6-10. In the load cases, F_S is replaced by F_T for the orthogonal wind.

Table 6-12. Total Force, Structural Frame and Appurtenances, Equipment, Piping, and Others, F_T (Wind Toward Frame 3).

Floor level	*Tributary height (ft)*	*Width (ft)*	A_g *(ft²)*	q_z *(psf)*	G	C_f	F *(lb)*
0	0–10 = 10	41	410	38.9	0.85	1.71	23,182
1	10–34 = 24	41	984	46.1	0.85	1.71	65,934
2	34–65 = 31	41	1,271	52.9	0.85	1.71	97,727
3	65–83 = 18	41	738	55.7	0.85	1.71	59,748
$F_T = \sum F =$							246,592

Note: To convert pound force (lb) to N, multiply *F* values in this table by 4.448.

Table 6-13. Total Force, Structural Frame and Appurtenances, Equipment, Piping, and Others, F_T (Wind Toward Frame A).

Floor level	*Tributary height (ft)*	*Width (ft)*	A_g *(ft²)*	q_z *(psf)*	G	C_f	F *(lb)*
0	0–10 = 10	50	500	38.9	0.85	1.60	26,452
1	10–34 = 24	50	1,200	46.1	0.85	1.60	75,235
2a	34–49 = 15	50	750	52.9	0.85	1.60	53,958
2b	49–65 = 16	41	656	52.9	0.85	1.60	47,195
3	65–83 = 18	41	738	55.7	0.85	1.60	55,905
$F_T = \sum F =$							258,745

Note: To convert pound force (lb) to N, multiply *F* values in this table by 4.448.

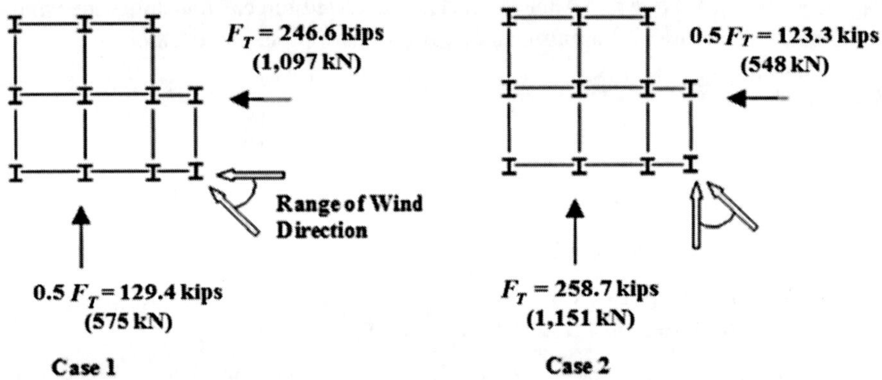

Figure 6-10. Design load cases for open-frame example (using method of Appendix 5B).

6.4 PARTIALLY CLAD STRUCTURE EXAMPLE

This section demonstrates the calculation of wind loads for partially clad structures per Section 5.4. For this example, it is assumed that the stair tower for the structure depicted in Figures 6-5, 6-6, and 6-7 is to be partially clad. Wind loads on the balance of the structure are not considered in this example, including any shielding effects, and therefore need to be calculated separately. Assume that the 9 ft wide face along Column Line A and beyond Column Line 3 is to be clad from elevation 0 to elevation 48 ft (the full height of the stair tower). Also assume that the 20 ft wide exterior face between Column Lines A and B is to be clad for the full height. For the purpose of this example, assume that north is up on the diagrams for the open-frame structure in Section 6.3.

In this example, the structure is classified as Risk Category III, and therefore the basic wind speed is 155 mi/h.

Design wind forces are determined by Equation (5-1a)

$$F = q_z G C_f A_e$$

The velocity pressure, q_z, is determined by ASCE 7, Equation (26.10-1) for the midheight of the stair tower

$$K_z = 0.94 \text{ for } h = 24 \text{ ft } (7.32 \text{ m})$$

$$q_z = 0.00256 K_z K_{zt} K_d K_e V^2 = (0.00256)(0.94)(1.0)(0.85)(1.0)(155)^2$$

$$= 49.0 \text{ psf } (2.36 \text{ kN/m}^2)$$

The gust effect factor, G, is equal to 0.85 per ASCE, Section 26.11 for rigid structures.

Force coefficients for a structure with two adjacent walls clad:

For wind directions for which the unclad faces are windward (wind from the north and west), $C_f = 2.0$ for these two directions simultaneously.

For wind directions for which the clad faces are windward (wind from the south and east), $C_f = 1.5$ for these two directions simultaneously.

The projected areas are the height times the width of each face.

For the face along Column Line A (wind from the north or south)

$$A_e = 48 \text{ ft} \times 9 \text{ ft} = 432 \text{ ft}^2 \ (40.1 \text{ m}^2)$$

For the face located between Column Lines A and B (wind from the east or west)

$$A_e = 48 \text{ ft} \times 20 \text{ ft} = 960 \text{ ft}^2 \ (89.1 \text{ m}^2)$$

Wind force for each wind direction:

Wind from north to south: $F = 49.0 \text{ psf} \times 0.85 \times 2.0 \times 432 \text{ ft}^2 = 35{,}986 \text{ lb}$ (161 kN)

Wind from east to west: $F = 49.0 \text{ psf} \times 0.85 \times 1.5 \times 960 \text{ ft}^2 = 59{,}976 \text{ lb}$ (268 kN)

Wind from south to north: $F = 49.0 \text{ psf} \times 0.85 \times 1.5 \times 432 \text{ ft}^2 = 26{,}989 \text{ lb}$ (121 kN)

Wind from west to east: $F = 49.0 \text{ psf} \times 0.85 \times 2.0 \times 960 \text{ ft}^2 = 79{,}968 \text{ lb}$ (357 kN)

The application of the forces is shown in Figure 6-11 (north is upward on the plan view). Two load cases are required to reflect the simultaneous application of north-west winds and south-east winds.

6.5 PRESSURE VESSELS EXAMPLE

This section will demonstrate the application of the recommended guidelines for calculating wind loads on pressure vessels. In this example the vessel is classified as Risk Category III and therefore the basic wind speed is 155 mi/h.

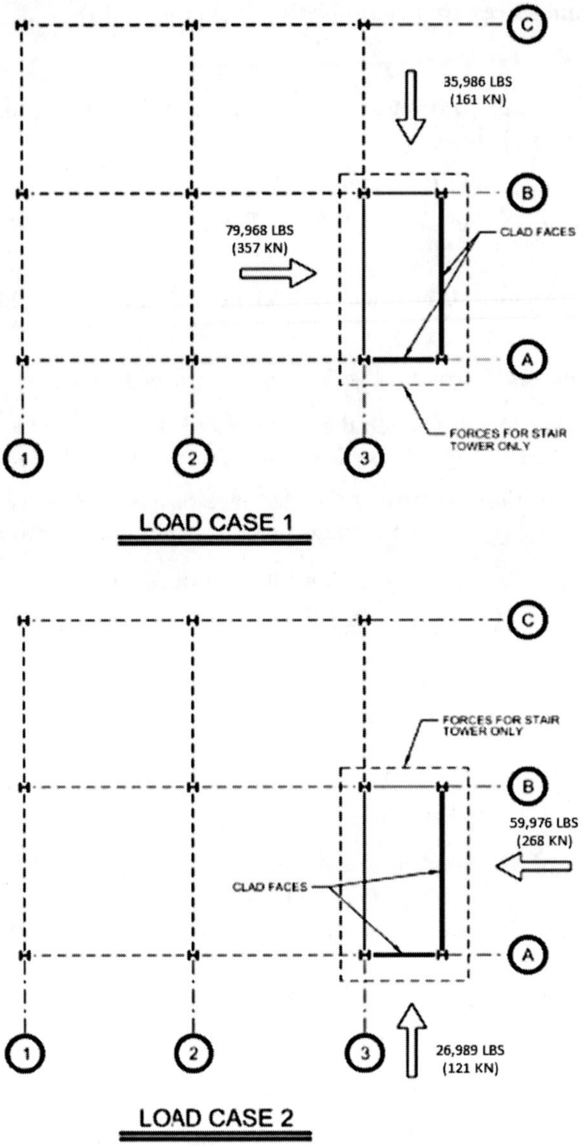

Figure 6-11. Partially clad structure load cases.

6.5.1 Vertical Vessel

Reference Figure 6-12 for a description of the ground-supported vessel used in this example. Assume the closest vertical vessel is spaced greater than three times the vessel diameter (as in Section 5.5.1.4).

6.5.1.1 *Simplified Method: Rigid Vessel*

Wind loads for rigid vessels using the simplified method determined by using Equation (5-1a) ($F = q_z G C_f A_e$) are shown in Table 6-14. The velocity pressure, q_z, is determined by $0.00256 \times K_z K_{zt} K_d K_e V^2$, where $K_d = 0.95$. Other terms in this equation are determined as follows from ASCE 7, Section 26.11:

$G = 0.85$
$h/D = 150/10 = 15$

The vessel is rough per Section 5.5.1.2; therefore, from ASCE 7, Figure 29.4-1

$C_f = 0.8 + 8 \times 0.1/18 = 0.84$ (by interpolation)

Increased diameter to approximate appurtenances (Section 5.5.1.2)

$D + 5$ ft $= 10 + 5 = 15$ ft or
$D + 3$ ft $+$ diameter of largest pipe $= 10 + 3 + 1.5 = 14.5$ ft
Largest value controls; therefore, effective $D = 15$ ft (4.57 m)

Therefore, $A_e = 15\Delta h$

6.5.1.2 *Detailed Method: Rigid Vessel*

Vessel and Miscellaneous

Wind loads for a rigid vessel for the detailed method determined by using Equation (5-1a) ($F = q_z C G_f A_e$) are given in Table 6-15. The velocity pressure, q_z, is determined by $0.00256 \times K_z K_{zt} K_d K_e V^2$, where $K_d = 1.0$. Other terms in the equation are determined as follows from ASCE 7, Section 26.11:

$G = 0.85$
$h/D = 150/10 = 15$

Table 6-14. Simplified Method: Calculation of Base Shear.

Height aboveground (ft)	K_Z	q_z *(psf)*	G	C_f	A_e *(ft²)*	F *(lb)*
0–15	0.85	49.7	0.85	0.84	225	7,984
15–20	0.90	52.6	0.85	0.84	75	2,817
20–40	1.04	60.8	0.85	0.84	300	13,023
40–60	1.13	66.0	0.85	0.84	300	14,137
60–80	1.21	70.7	0.85	0.84	300	15,144
80–100	1.26	73.6	0.85	0.84	300	15,765
100–120	1.31	76.5	0.85	0.84	300	16,386
120–140	1.36	79.5	0.85	0.84	300	17,029
140–160	1.39	81.2	0.85	0.84	300	17,393
Total = 119,679 lb (532 kN)						

Table 6-15. Detailed Method: Vessel and Miscellaneous, Calculation of Base Shear.

Height aboveground (ft)	K_Z	q_z *(psf)*	G	C_f	A_e *(ft²)*	F *(lb)*
0–15	0.85	52.3	0.85	0.64	172.5	4,908
15–20	0.90	55.4	0.85	0.64	57.5	1,733
20–40	1.04	64.0	0.85	0.64	230	8,008
40–60	1.13	69.3	0.85	0.64	230	8,671
60–80	1.21	74.4	0.85	0.64	230	9,309
80–100	1.26	77.5	0.85	0.64	230	9,697
100–120	1.31	80.6	0.85	0.64	230	10,085
120–140	1.36	83.6	0.85	0.64	230	10,460
140–150	1.39	85.5	0.85	0.64	115	5,349
				Total = 68,219 lb (304 kN)		

Per Section 5.5, the vessel is determined as moderately smooth; therefore, from ASCE 7, Figure 29.4-1

$$C_f = 0.6 + 8 \times 0.1/18 = 0.64$$

Diameter is increased to approximate ladder, nozzles, and piping 8 in. or smaller, Section 5.5.1.3 of this report

$$D + 1.5 \text{ ft} = 10 + 1.5 \text{ ft (3.51 m)}$$

Therefore, $A_e = 11.5\Delta h$

Large-Diameter Pipe (>8 in.)

Wind loads for piping determined using Equation (5-1a), $F = q_z G C_f A_e$, are given in Table 6-16. The velocity pressure, q_z, is determined by $0.00256 \times K_z K_{zt} K_d K_e V^2$, where $K_d = 1.0$. Other terms in equation are determined as follows:

$G = 0.85$ from ASCE 7, Section 26.11
$C_f = 0.7$ from Section 5.5.1.3
Pipe diameter = 18 in. = 1.5 ft (0.46 m)

Therefore, $A_e = 1.5\Delta h$

A_e above elevation 150 = $3.14 \times 10/2 \times 1.5 = 24$ ft² (2.2 m²)

[The pipe starts at elevation 15.0 (Figure 6-12)]

Platforms

Refer to Figure 6-12.

The platform at elevation 150 (just above the top of the vessel) is a square platform 12 ft × 12 ft in plan with a handrail around the perimeter. It is assumed that the platform structural framing will be 8 in. deep (8 in./12 in./ft = 0.70 ft²/ft). The critical case will be one in which the wind direction is diagonal to the square

Table 6-16. Detailed Method Pipe: Calculation of Base Shear.

Height aboveground (ft)	K_Z	q_z *(psf)*	G	C_f	A_e *(ft²)*	F *(lb)*
15–20	0.90	55.4	0.85	0.70	7.5	247
20–40	1.04	64.0	0.85	0.70	30	1,142
40–60	1.13	69.5	0.85	0.70	30	1,237
60–80	1.21	74.4	0.85	0.70	30	1,328
80–100	1.26	77.5	0.85	0.70	30	1,383
100–120	1.31	80.6	0.85	0.70	30	1,439
120–140	1.36	83.6	0.85	0.70	30	1,492
140–150	1.39	85.5	0.85	0.70	15	763
150–155	1.39	85.5	0.85	0.70	24	1,221
						Total = 10,253 lb (45.6 kN)

platform; therefore, the length of the platform will be multiplied by the square root of 2 (i.e., 1.414).

Therefore, A_e (platform framing) = 0.7 × 12 × 1.414 = 11.88 ft²

A_e (front handrail) = 0.8 × 12 × 1.414 = 13.58 ft²
A_e (back handrail) = 0.8 × 12 × 1.414 = 13.58 ft² from Section 5.5.1.3
39.03 ft² (3.62 m²)

$q_z = 0.00256 \times K_z \times K_{zt} \times K_d \times K_e \times V^2 = 72.7$ psf
$K_d = 0.85$
$G = 0.85$ from ASCE 7, Section 26.11
$C_f = 2.0$ from Section 5.5.1.3
$F = q_z G C_f A_e = 72.7 \times 0.85 \times 2.0 \times 39.03 = 4{,}824$ lb (21.5 kN)

The other platforms are circular and extend 3 ft beyond the outside radius of the vessel. Therefore, the radial distance (R) from the centerline of the vessel to the outside of the platform is 5 + 3 = 8 ft (2.44 m). The angle (60°, 90°, and 180°) as shown in Figure 6-12 is the angle subtended by the ends of the platform as measured at the centerline of the vessel. Therefore, the projected length of the platform is calculated by the equation

$L = 2R\sin(\text{subtended angle}/2)$

Platform at elevation 100 ft. – subtended angle 60°
Projected length = 2 × 8 × sin(60/2) = 8.0 ft
Assume platform framing is 6 in. deep (0.5 ft²/linear ft)

A_e (platform framing) = 0.5 × 8.0 = 4.0 ft²
A_e (handrail) = 0.8 × 8.0 = 6.4 ft²
10.4 ft² (0.97 m²)
$q_z = 0.00256 \times 1.26 \times 0.85 \times (155)^2 = 65.9$ psf
$G = 0.85$ from ASCE 7, Section 26.11

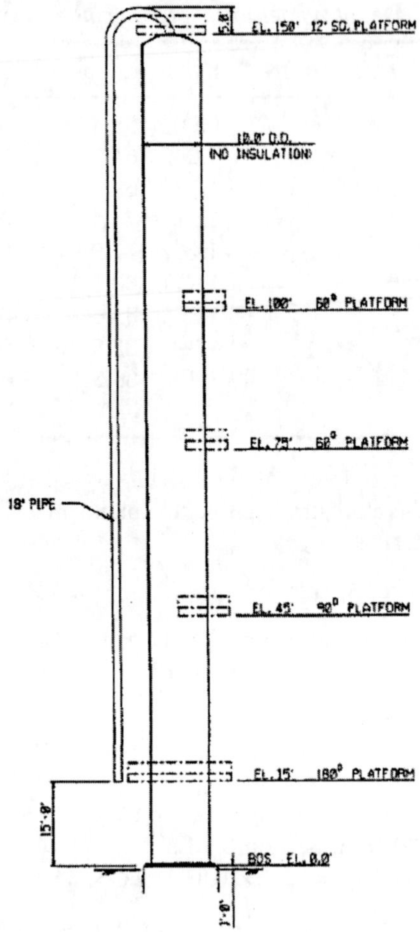

Shell thickness t = 1"
Empty Weight = 280 kips
Operating Weight = 500 kips

Figure 6-12. Vertical vessel example.

C_f = 2.0 from Section 5.5.1.3
$F = q_zGC_fA_e = 63.3 \times 0.85 \times 2.0 \times 10.4 = 1{,}119$ lb (5.0 kN)

Platform at elevation 45 ft, subtended angle = 90°

Projected length = $2 \times 8 \times \sin(90°/2) = 11.3$ ft. Because the projected length is larger than the vessel diameter, the back handrail will be included (Section 5.5.1.3). Back handrail project length = 3 ft × sin (45°) × 2 sides = 4.2 ft.

Assume platform framing is 6 in. deep (0.5 ft^2/ft)

A_e (platform framing) = 0.5 × 11.3 = 5.7 ft^2
A_e (front handrail) = 0.8 × 11.3 = 9.0 ft^2
A_e (back handrail) = 0.8 × 4.2 = 3.3 ft^2 from Section 5.5.1.3
18.0 ft^2 (1.67 m^2)

$q_z = 0.00256 \times 1.13 \times 0.85 \times (155)^2 = 59.1$ psf
$G = 0.85$ from ASCE 7, Section 26.11
$C_f = 2.0$ from Section 5.5.1.3

$F = q_z G C_f A_e = 59.1 \times 0.85 \times 2.0 \times 18.0 = 1{,}808$ lb (8.0 kN)

Platform at elevation 15 ft, subtended angle = 180°

Projected length = 2 × 8 × sin(180/2) = 16 ft. Because the projected length is larger than the vessel diameter, the back handrail will be included (Section 5.5.1.3). Back handrail project length = 3 ft × sin90° × 2 sides = 6 ft.

Assume platform framing is 6 in. deep (0.5 ft^2/ft)

A_e (platform framing) = 0.5 × 16.0 = 8.0 ft^2
A_e (front handrail) = 0.8 × 16.0 = 12.8 ft^2
A_e (back handrail) = 0.8 × 6.0 = 4.8 ft^2 from Section 5.5.1.3
25.6 ft^2 (2.38 m^2)

$q_z = 0.00256 \times 0.85 \times 0.85 \times (155)^2 = 44.4$ psf
$G = 0.85$ from ASCE 7, Section 26.11
$C_f = 2.0$ from Section 5.5.1.3

$F = q_z G C_f A_e = 44.4 \times 0.85 \times 2.0 \times 25.6 = 1{,}932$ lb (8.6 kN)

Total shear on platforms = 4,824 + 1,165 + 1,119 + 1,808 + 1,932 = 10,849 lb (48.3 kN)

Total shear on foundation:

Vessel and miscellaneous	= 68,219 lb
Large-diameter pipe (>8 in.)	= 10,253 lb
Platforms	= 10,849 lb
Total	89,320 lb (397 kN)

6.5.1.3 Analysis of Flexible Vessels

The only difference between loads resulting from the analysis of the vessel as "flexible" versus "rigid" is that G_f (the gust effect factors for main wind-force-resisting systems of flexible buildings and structures) is substituted for G in the rigid analysis.

For the recommended methods, the procedure outlined in ASCE 7, Section 6.5.8.2 was utilized to determine $G_f = 1.099$:

- Simplified (flexible):

  Total shear = $(G_f/G) \times 119{,}679$ lb
  $= (1.198/0.85) \times 119{,}679 = 168{,}677$ lb (750 kN)

- Detailed (flexible):

  Vessel and miscellaneous:
  Shear = $(G_f/G) \times 68{,}219$ lb = $(1.198 / 0.85) \times 68{,}219 = 96{,}148$ lb

  Pipe:
  Shear = $(G_f/G) \times 10{,}253$ lb = $(1.198/0.85) \times 10{,}253 = 14{,}451$ lb

  Platforms:
  Shear = $(G_f/G) \times 10{,}849$ lb = $(1.198/0.85) \times 10{,}849 = 15{,}291$ lb
  Total = 125,890 lb (560 kN)

Gust Effect Factor per ASCE 7 (Customary Units)

Design parameters:
Wind speed: $V = 155$ mi/h
Topographic factor: $K_{zt} = 1.0$

Exposure category: C ASCE Table 26.11-1

$$\bar{b}=0.65 \qquad c=0.20 \qquad l=500 \qquad \bar{\epsilon}=\frac{1}{5} \qquad \bar{\alpha}=\frac{1}{6.5}$$

$$Z_g=900 \qquad \bar{\alpha}=\frac{1}{9.5} \qquad Z_{min}=15$$

Vessel diameter: $D = 10$ ft (excluding insulation)
Vessel height: $h = 150$ ft

First, consider the empty case:

Percentage increase of vessel empty weight: 1.1×280 kips = 28.0 kips (add 10% to account for piping and platform)
Vessel empty weight: $W = (280 \text{ kips} + 28 \text{ kips})/150 \text{ ft} = 2{,}053$ lb/ft
Shell thickness: $t = 1.0$ in.
Damping ratio: $\beta = 0.01$ (as commonly assumed value)

Vessel period:

$$T=\frac{7.78}{10^6}\left(\frac{H}{D}\right)^2\sqrt{\frac{12WD}{t}} \qquad T=0.869\text{ s}$$

Vessel frequency: $n_1 = 1/T$ $\qquad n_1 = 1.151\text{ s}^{-1}$

The frequency is greater than 1 Hz and is considered rigid for empty condition; therefore, use $G_f = 0.85$ per ASCE 7, Section 26.11.1.

Now consider the operating case:

Vessel operating weight: W = (500 kips + 28.0 kips)/150 ft = 3,520 lb/ft
Shell thickness: t = 1.0 in.
Damping ratio: β = 0.01

Vessel period:

$$T = \frac{7.78}{10^6}\left(\frac{H}{D}\right)^2\sqrt{\frac{12WD}{t}} \qquad T = 1.138\ \mathrm{s}$$

Vessel frequency: $n_1 = 1/T$ $\qquad n_1 = 0.\ 879\ \mathrm{s}^{-1}$

The frequency is less than 1 Hz, so the vessel is considered flexible for the operating condition. The value of G_f must be calculated according to ASCE 7, Section 26.11.5.

Gust effect factor (equations from ASCE 7, Section 26.11.5)

For flexible structures:

$$G_f = 0.925\left(\frac{1 + 1.7I_{\bar{Z}}\sqrt{g_Q^2Q^2 + g_R^2R^2}}{1 + 1.7g_V I_{\bar{Z}}}\right) \qquad (26.11\text{-}10)$$

g_Q and g_V shall be taken as 3.4

$$g_R = \sqrt{2\ln\left(3{,}600\eta_1\right)} + \frac{0.577}{\sqrt{2\ln(3{,}600\eta_1)}} g_R = 4.159 \qquad (26.11\text{-}11)$$

R, the resonant response factor, is given by

$$R = \sqrt{\frac{1}{\beta}R_n R_h R_B(0.53 + 0.47R_L)} \qquad (26.11\text{-}12)$$

$$R_n = \frac{7.47N_1}{(1 + 10.3N_1)^{5/3}} R_n = 0.068 \qquad (26.11\text{-}13)$$

$$\text{where} \quad N_1 = \frac{\eta_1 L_{\bar{Z}}}{V_{\bar{Z}}} N_1 = 3.11 \qquad (26.11\text{-}14)$$

$$\text{and} \quad \overline{V}_{\overline{Z}} = \overline{b}\left(\frac{\overline{Z}}{33}\right)^{\overline{\alpha}} V\left(\frac{88}{60}\right) \quad \overline{V}_{\overline{Z}} = 172.43\,\text{ft/s} \tag{26.11-16}$$

$$R_l = \frac{1}{\eta} - \frac{1}{2\eta^2}\left(1 - e^{-2\eta}\right) \tag{26.11-15a}$$

where $R_l = R_h$ setting $\eta = \frac{4.6\eta_1 h}{\overline{V}_{\overline{Z}}} = 3.517$ $R_h = 0.244$

$R_l = R_B$ setting $\eta = \frac{4.6\eta_1 B}{\overline{V}_{\overline{Z}}} = 0.234$ $R_B = 0.861$, where $B = 10$

$R_l = R_L$ setting $\eta = \frac{15.4\eta_1 L}{\overline{V}_{\overline{Z}}} = 0.785$ $R_L = 0.631$, where $L = 10$

Substituting these values into ASCE 7 Equation (26.11-12), the resonant response factor $R = 1.086$.

From ASCE 7, the intensity of turbulence at height $\overline{Z}$

$$I_{\overline{Z}} = c\left(\frac{33}{\overline{Z}}\right)^{1/6} = 0.169 \tag{26.11-7}$$

where $\overline{Z} = 0.6h \geq Z_{\min}$.

From ASCE 7, the background response Q is given by

$$Q = \sqrt{\frac{1}{1 + 0.63\left(\frac{B+h}{L_{\overline{Z}}}\right)^{0.63}}} \qquad Q = 0.887 \tag{26.11-8}$$

where $L_{\overline{Z}} = l(\overline{Z}/33)^{\overline{\epsilon}} = 611$ in which l and $\overline{\epsilon}$ are listed in Table 26.11-1 of ASCE 7.

Now, substituting all the variables into ASCE 7, Equation (26.11-10)

$$G_f = 0.925\left(\frac{1 + (1.7 \times 0.169)\sqrt{3.4^2 \times 0.887^2 + 4.159^2 \times 1.086^2}}{1 + 1.7 \times 3.4 \times 0.169}\right)$$

$$G_f = 1.198$$

Note that G_f was determined to be 1.099 in the previous version of this book (ASCE 2011), which was based on ASCE 7-05 (2005). The increase in the computation of G_f here is because of the increase in basic wind speed in ASCE 7-16 compared to ASCE 7-05. The site location and computation methodology and formulas to determine G_f remain the same.

The committee noticed during the preparation of this example that there can be significant changes in the computation of the gust effect factor at the threshold between rigid and flexible structure classifications. Judgment is required in these cases. As noted in Section 2.7, this should be a topic for future research.

6.5.2 Horizontal Vessels

For description of the horizontal vessel used in this example, see Figure 6-13.

6.5.2.1 Transverse Wind (Wind on the Side of the Vessel)

$F = q_z G C_f A_e$, A_e is defined in Section 5.5.2.2

$h = 20$ ft at platform level, $K_z = 0.90$, $K_d = 0.85$, and $q_z = 0.00256 \times 0.85 \times 0.90 \times 155^2 = 47.1$ psf

Vessel and Miscellaneous

$G = 0.85$ from ASCE 7, Section 26.1111.1
$B/D = 50/12 = 4.2$, assuming vessel is moderately smooth

Therefore,

$C_f = 0.5 + 3.2 \times 0.1/6 = 0.55$ (by interpolation) from ASCE 7, Figure 29.4-1

Increased diameter to approximate ladder, nozzles, and piping 8 in. or smaller:

$D + 1.5$ ft $= 12 + 1.5 = 13.5$ ft (4.1 m) from Section 5.5.2.2

Therefore, $A_e = 13.5 \times 54$ (avg.) $= 729$ ft^2 (67.7 m^2)

Calculate shear at base

$F = q_z G C_f A_e = 47.1 \times 0.85 \times 0.55 \times 729 = 16{,}052$ lb (71.4 kN)

Platform

The platform is a rectangular platform 10 ft × 30 ft in plan with a handrail around the perimeter. It is assumed that the platform structural framing will be 10 in. deep (10 in./12 in./ft = 0.8 ft^2/ft).

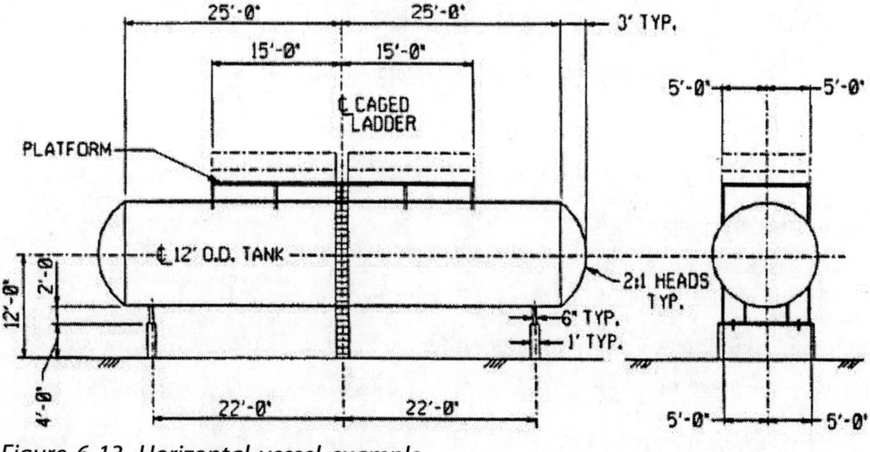

Figure 6-13. Horizontal vessel example.

Therefore, A_e (platform framing) $= 0.8 \times 30 = 24.0$ ft^2
A_e (front handrail) $= 0.8 \times 30 = 24.0$ ft^2
A_e (back handrail) $= 0.8 \times 30 = \underline{24.0 \text{ ft}^2}$ from Section 5.5.2.6
72.0 ft^2 (6.7 m^2)

$q_z = 47.1$ psf
$G = 0.85$ from ASCE 7, Section 26.11
$C_f = 2.0$ from Section 5.5.2.6

$$F = q_z G C_f A_e = 47.1 \times 0.85 \times 2.0 \times 72.0 = 5{,}765 \text{ lb } (25.6 \text{ kN})$$

Supports

Steel saddle:

$A_e = 0.5 \times 2.0 \times 2$ (supports) $= 2.0$ ft^2 (0.2 m^2)
$q_z = 47.1$ psf
$G = 0.85$ from ASCE 7, Section 26.11
$C_f = 2.0$ from Section 5.5.2.6

$$F = q_z GCfA_e = 47.1 \times 0.85 \times 2.0 \times 2.0 = 160 \text{ lb } (0.7 \text{ kN})$$

Concrete support:

$A_e = 1.0 \times 4.0 \times 2$ (supports) $= 8.0$ ft^2 (0.7 m^2)
$q_z = 47.1$ psf
$G = 0.85$ from ASCE 7, Section 26.11
$C_f = 1.3$ from Section 5.5.2.7

$$F = q_z G C_f A_e = 47.1 \times 0.85 \times 1.3 \times 8.0 = 416 \text{ lb } (1.9 \text{ kN})$$

Total Shear, Transverse Wind

Vessel and miscellaneous	= 16,052 lb
Platform	= 5,765 lb
Supports (steel)	= 160 lb
Supports (concrete)	= 416 lb
Total	22,394 lb (99.6 kN)

6.5.2.2 Longitudinal Wind (Wind on the End of Vessel)

$$F = q_z G C_f A_e$$

$h = 20$ ft at platform level, $q_z = 47.1$ psf

Vessel and Miscellaneous

$G = 0.85$ from ASCE 7, Section 26.11
Elliptical head; therefore, $C_f = 0.5$ from Section 5.5.2.4

Increased diameter to approximate ladder, nozzles, and piping 8 in. or smaller

$$D + 1.5 \text{ ft} = 12 + 1.5 = 13.5 \text{ ft } (4.1 \text{ m}) \text{ from Section 5.5.2.2}$$

Therefore, $A_e = \pi \times 13.5 \times 13.5/4 = 143.1$ ft^2 (13.3 m^2)

Calculate shear at base:

$F = q_z G C_f A_e = 47.1 \times 0.85 \times 0.5 \times 143.1 = 2{,}865$ lb (12.7 kN)

Platform

The platform is a rectangular platform 10 ft × 30 ft in plan with a handrail around the perimeter. It is assumed that the platform structural framing will be 10 in. deep (10 in./12 in. = 0.8 ft²/ft).

Therefore, A_e (platform framing) = $0.8 \times 10 = 8.0$ ft²
A_e (front handrail) = $0.8 \times 10 = 8.0$ ft² from Table 5-1
A_e (back handrail) = $0.8 \times 10 = \underline{8.0}$ ft² from Section 5.5.2.6
24.0 ft² (2.2 m²)

$q_z = 47.1$ psf
$G = 0.85$ from ASCE 7, Section 26.11
$C_f = 2.0$ from Section 5.5.2.6
$F = q_z G C_f A_e = 47.1 \times 0.85 \times 2.0 \times 24.0 = 1{,}922$ lb (8.6 kN)

Supports

Steel saddle: 10 ft wide × 3 ft (avg.) high
$A_e = 10.0 \times 3.0 \times 2$ (supports) = 60.0 ft² (5.6 m²)
$q_z = 47.1$ psf
$G = 0.85$ from ASCE 7, Section 26.11
$C_f = 2.0$ from Section 5.5.2.6
$F = q_z G C_f A_e = 47.1 \times 0.85 \times 2.0 \times 60.0 = 4{,}804$ lb (21.4 kN)

Concrete support: 11 ft wide × 4 ft high
$A_e = 11.0 \times 4.0 \times 2$ (supports) = 88.0 ft² (8.2 m²)
$q_z = 47.1$ psf
$G = 0.85$ from ASCE 7, Section 26.11
$C_f = 1.3$ from Section 5.5.2.7
$F = q_z G C_f A_e = 47.1 \times 0.85 \times 1.3 \times 88.0 = 4{,}580$ lb (20.4 kN)

Total Shear, Longitudinal Wind

Vessel and miscellaneous	= 2,865 lb
Platforms	= 1,922 lb
Supports (steel)	= 4,804 lb
Supports (concrete)	= 4,580 lb
Total	14,171 lb (63.0 kN)

6.6 COOLING TOWER EXAMPLE

A wood-frame, dual-cell, cooling tower with metal siding and metal louvers is located at the industrial site near the Gulf Coast described in Section 6.1.

Figure 6-14. Cooling tower for example problem.

The tower is 32 ft long × 20 ft wide × 12 ft high. The fan cylinders are 6 ft high and 8 ft in diameter. Each metal siding panel is 2 ft wide × 8 ft long. The metal louvers are 3 ft long and 1 ft wide, with an inward slope of 45°. The cooling tower is shown in Figure 6-14. The fan deck flooring (i.e., roof) is 2 in. × 6 in. tongue-and-groove wooden planking material. The site is in a rural area. The tower lies along the edge of the plant, bordering a flat, open area that extends for half a mile. The basic wind speed per ASCE 7 is 145 mi/h. The cooling tower services various administrative support facilities at the plant, and therefore the structure is in Risk Category II.

Using Method 1, the simplified procedure of ASCE 7, for a mean roof height of 15 ft and assuming Exposure C, the adjustment factor λ will equal 1.21, with a topographic factor $K_{zt} = 1.0$ and a roof angle of 0°, the MWFRS can be calculated from ASCE 7 as follows:

$$p_s = \lambda K_{zt} p_{s30} \tag{28.5-1}$$

where p_{s30} is from ASCE 7, Figure 28.5-1.

Horizontal load on walls:

Zone A: 10% × 20 ft = 2 ft < 0.4 × 12 ft = 4.8 ft (1.46 m)
2 ft < 3 ft minimum, so a = 3 ft (0.91 m)

Use $2a = 2 \times 3$ ft = 6 ft. The edge zone is from corners along both sides (1.21) (1.0) (33.5 psf) = 40.5 psf (0.18 kN/m²)

Note: The value of 33.5 psf is an interpolation of the wind pressure between the wind speed values of 140 and 150 mi/h at 0°–5°.

Zone C: 32 ft – 2 (6 ft) = 20 ft. The interior zone is on the long side (6.1 m)
20 ft – 2 (6 ft) = 8 ft. The interior zone is on the short side (2.44 m)
(1.21) (1.0) (22.2 psf) = 26.8 psf (0.12 kN/m^2)

Vertical load on roof:

Zone E: Use $2a$ = 6 ft. The edge zone is around the perimeter of the roof (1.836 m)
(1.21) (1.0) (−40.2 psf) = −48.6 psf (uplift) (−0.22 kN/m^2)

Zone G: 32 ft – 2 (6 ft) = 20 ft. The interior zone is on the long side (6.1 m)
20 ft – 2 (6 ft) = 8 ft. The interior zone on the short side (2.43 m)
(1.21) (1.0) (−27.9 psf) = −33.8 psf (uplift) (−0.1215 kN/m^2)

Note: A 20 ft × 8 ft interior zone (Zone G) will coincide with fan openings.

The simplified procedure of ASCE 7 designates two edge zones, E and F, with Zone E being the greater value. The procedure also designates two interior zones, G and H, with Zone G being the greater value. To simplify analysis of design wind loads on the MWFRS, the higher Zone E value is used for the edge zone, while the higher Zone G value is used for the interior zone. In most cases, the Zone G area will coincide with the fan openings.

Components and cladding can be calculated from ASCE 7 as follows:

$$p_{net} = \lambda K_{zt} p_{net30} \qquad (30.4\text{-}1)$$

where p_{net30} is from ASCE 7, Figure 30.4-1.

Metal cladding panel:

Effective wind area: 2 ft × 8 ft = 16 ft^2 > 10 ft^2, use 16 ft^2 (1.49 m^2)

Zone 4: 10% × 20 ft = 2 ft < 0.4 × 12 ft = 4.8 ft (1.46 m)
(Interior of wall) 2 ft < 3 ft minimum, so a = 3 ft (0.91 m)
20 ft – 2 (3 ft) = 14 ft. The interior zone is on the short side (4.27 m)
(1.21) (1.0) (36.9 psf) = 44.6 psf (inward) (0.20 kN/m^2)
(1.21) (1.0) (−37.4 psf) = −45.3 psf (outward) (0.20 kN/m^2)

Zone 5: a = 3 ft. The end zone is from the corners along the short side (0.91 m)
(End of wall) (1.21) (1.0) (36.9 psf) = 44.6 psf (inward) (0.20 kN/m^2)
(1.21) (1.0) (−45.2 psf) = −57.4 psf (outward) (0.24 kN/m^2)

Metal louver:
Effective wind area: 3 ft × 1 ft = 3.0 ft^2 < 10 ft^2, use 10 ft^2 (0.93 m^2)

Zone 4: 10% × 20 ft = 2 ft < 0.4 × 12 ft = 4.8 ft (1.46 m)
(Interior of wall) 2 ft < 3 ft minimum, so a = 3 ft (0.91 m)
32 ft – 2 (3 ft) = 26 ft. The interior zone is on the long side (7.92 m)
(1.21) (1.0) (37.5 psf) = 45.4 psf (inward) (0.20 kN/m^2)
(1.21) (1.0) (−38 psf) = −46 psf (outward) (0.20 kN/m^2)

Zone 5: a = 3 ft. The end zone is from the corners along the long side (0.91 m)
(End of wall) (1.21) (1.0) (37.5 psf) = 45.4 psf (inward) (0.20 kN/m^2)
(1.21) (1.0) (−47 psf) = −56.9 psf (outward) (0.25 kN/m^2)

Although the metal louver is inwardly inclined, air flow between the louvers is restricted, which could cause an increase in local pressure normal to the louver surface. Moreover, separation of air flow along the leading edge of a louver may redirect and accelerate air flow toward an adjacent louver; hence, the total surface area of the louver is used without adjusting for the angle of inclination.

Uplift load on deck planks:
Zone 1′ in ASCE 7 was not considered in this example.

Effective wind area: 10 ft × 6 in./12 in./ft = 5.0 ft^2 < 10 ft^2, use 10 ft^2 (0.93 m^2)
Zone 1: 10% × 20 ft = 2 ft < 0.4 × 12 ft = 4.8 ft (1.46 m)
(Interior of roof deck) 2 ft < 3 ft minimum, so a = 3 ft (0.91 m)
32 ft – 2 (3 ft) = 26 ft. The interior zone is on the long side (7.92 m)
20 ft – 2 (3 ft) = 14 ft. The interior zone is on the short side (4.27 m)
(1.21) (1.0) (15 psf) = 18.2 psf (inward) (0.08 kN/m^2)
(1.21) (1.0) (−56 psf) = −67.8 psf (outward) (0.30 kN/m^2)

Zone 2: a = 3 ft. The end zone is on the long and short sides (0.91 m)
(Ends of roof deck) (1.21) (1.0) (15 psf) = 18.2 psf (inward) (0.08 kN/m^2)
(1.21) (1.0) (−74 psf) = −89.5 psf (outward) (0.40 kN/m^2)

Zone 3: $a = 3$ ft. The square zone is at all four corners (0.91 m)
(Corners of roof deck) (1.21) (1.0) (15 psf) = 18.2 psf (inward)
(0.08 kN/m²)

(1.21) (1.0) (−101 psf) = −122.2 psf (outward)
(0.54 kN/m²)

Horizontal load on fan cylinder:

$$p_z = K_z\ K_{zt}\ (V/28)^2 \qquad (5\text{-}9)$$

Projected area = (6 ft) (8 ft) = 48 ft² (4.46 m²)

$K_z = 0.88$, assuming Exposure Category C at height = 12 ft + 6 ft = 18 ft (5.49 m)

$p_z = (0.88)\ (1.0)\ (145/28)^2 = 23.6$ psf across projected area (0.1 kN/m²)

Horizontal load on handrail: Follow method as described in Section 5.3.

6.7 AIR COOLER (FIN-FAN) EXAMPLE

Assume the air cooler geometry is as follows (Figure 6-15):

- Single air cooler: 12 ft (3.6 m) wide and 24 ft (7.2 m) long and 4 ft (1.2 m) high.
- The bottoms of the air coolers (fin-fans) are assumed to be 9 ft (2.7 m) above the maintenance platform, which is 5 ft (1.5 m) above the top pipe level (elevation 35 ft) with 4 ft clearance underneath.
- Owing to the varying shapes, configurations, and attachments, a solid shape will be assumed that matches the maximum dimensions of the equipment.

Design wind force on the air cooler is determined by Equation (5-1a)

$$F = q_z G C_f A_e$$

Design wind pressure $q_z = 0.00256\ K_z K_{zt} K_d K_e V^2$ (lb/ft.²) from ASCE 7, Section 26.10.2

$$= 0.613\ K_z K_{zt} K_d K_e V^2\ (\text{N/m}^2)$$

From Section 6.1, $V = 145$ mi/h, and the exposure category is C.
Center of air cooler at elevation 46 ft ($K_z = 1.07$)

$$q_{46} = 49.2\text{ psf }(2.37\text{ kN/m}^2)$$

For the wind direction perpendicular (transverse) to the pipe rack, the projected area for each air cooler is

$$A_e = 12\text{ ft wide} \times 4\text{ ft high} = 48\text{ ft}^2\ (4.5\text{ m}^2)$$

From Table 5-3 $\quad a/b = 24$ ft/12 ft = 2.0 plan
$c/b = 4$ ft/12 ft = 0.33 elevation
$C_f = 0.91$

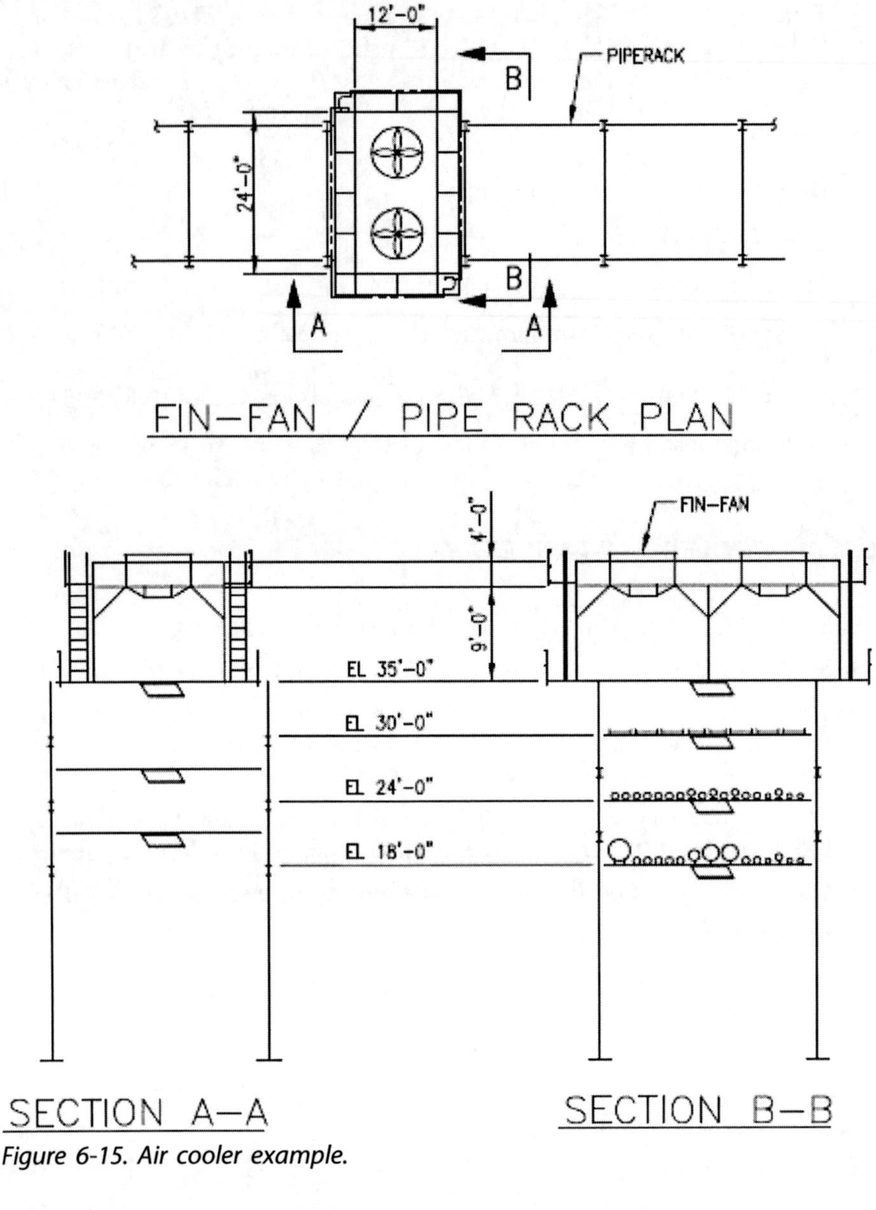

Figure 6-15. Air cooler example.

The force on the air cooler is

$F = 49.2 \text{ psf} \times 0.85 \times 0.91 \times 48 \text{ ft}^2 = 1{,}827 \text{ lb } (8.1 \text{ kN})$

For the wind direction parallel to the pipe rack, the projected area for each air cooler is

$A_e = 24 \text{ ft wide} \times 4 \text{ ft high} = 96 \text{ ft}^2\ (8.9 \text{ m}^2)$

From Table 5-3 $a/b = 12\ \text{ft}/24\ \text{ft} = 0.5$ plan
$c/b = 4\ \text{ft}/24\ \text{ft} = 0.167$ elevation
$C_f = 1.0$

The force on the air cooler is

$$F = 49.2\ \text{psf} \times 0.85 \times 1.0 \times 96\ \text{ft}^2 = 4{,}015\ \text{lb}\ (17.9\ \text{kN})$$

For wind loads at access platforms and pipe rack, see the open-frame examples.

References

ASCE. 2005. *Minimum design loads for buildings and other structures.* ASCE 7-05. Reston, VA: ASCE.

ASCE. 2011. *Wind loads for petrochemical and other industrial facilities.* Reston, VA: ASCE.

ASCE. 2017. *Minimum design loads and associated criteria for buildings and other structures.* ASCE 7-16. Reston, VA: ASCE.

Index

Figures are indicated by *f*; tables are indicated by *t*.

aerodynamics: crosswind forces, 10–11; of cylinders, 14; of open-frame structures, 11–12; of partially-clad structures, 13–14; of vertical vessels, 14–17
air coolers: application of wind load guidelines, 165–167, 166*f*; design wind load guidelines, 107–108, 110*t*
along-wind loads: on open-frame structures, 134, 136–137, 138*t*–139*t*, 139–141, 140*t*, 141*t*; on steel stacks, 27–28
appurtenances: aerodynamics of, 16; force coefficients and, 82–83, 83*t*
aspect ratio, 15–16, 117–118, 120*f*

building load combinations, 49, 50*t*, 51*t*, 52

cable trays: design wind load guidelines, 79, 80; pipe bridges, 131; pipe rack structures, 127
CFD (computational fluid dynamics) modeling, 47–48
cladding. *see* components and cladding
Code of Federal Regulations on LNG facilities, 63–68
components and cladding: cooling towers, 29–30, 163–164; force coefficients, 82–83, 83*t*; partially-clad structure aerodynamics, 13–14; performance during hurricanes, 38, 42*f*–43*f*, 44
compressor shelters, 109
computational fluid dynamics (CFD) modeling, 47–48
construction loads, 53–54
Cooling Technology Institute (CTI) Standard STD-114, 29
cooling towers: application of wind load guidelines, 161–165, 162*f*; design wind load guidelines, 106–107; existing design practices, 28–30; performance during hurricanes, 38, 41*f*–42*f*
cranes, 44, 45*f*
cross-wind forces: loading and, 10–11; open-frame structures, 141, 142*t*, 143, 143*t*–144*t*; steel stacks and, 28
cylinder aerodynamics, 14–16

Design and Construction of Large, Welded, Low-Pressure Storage Tanks, API 620 (2013), 26
design wind loads. *see* wind loads
directionality: vertical vessels and, 16–17; wind load formulation, 7
directional procedure, 2

electrical power transmission structure performance during hurricanes, 36, 38, 39*f*–41*f*
empty loads, 54
envelope procedure, 2
existing design practices: ASCE 7-16 and, 49; for cooling towers, 28–30; IBC and, 49; PIP STC01015 and, 49, 50*t*–51*t*, 52; for steel stacks,

27–28; survey of, 21–22, 23*t*–25*t*; for tanks, 22, 25–27
existing structures: design basis review, 72; friction loads, 62; hurricane event considerations, 57; international building codes, 70–71; live loads, 62; load combinations, 57, 58*t*–59*t*, 60*t*–61*t*, 62; load definitions, 52–54; modifications to, 57, 69–73; operational gravity loads, 57; past modifications to, 72–73; PIP STC01015 design criteria, 69–70; potential deterioration of, 73, 74*f*; supplemental guidelines for, 52–62; wind exposure category, 71–72, 73*f*; wind load evaluation, 58*t*–59*t*, 60*t*–61*t*, 62, 69–73

Federal Energy Regulatory Commission (FERC), 62–63
fin-fans. *see* air coolers
flare towers, 44, 44*f*
flexible vessels, 155–158
flooring, 88–89
force balance models, 46
force coefficients: alternate method for open-frame structures, 111–113, 113*f*, 115; cable trays and, 80; components and cladding and, 82–83, 83*t*; high-solidity open-frame structures and, 117–119, 120*f*; open-frame structures and, 84–86, 86*f*–87*f*, 111–113, 113*f*, 115, 134–137; partially-clad structures and, 149; pipes and, 79–80; structural members and, 79
friction loads, 62

gust effect: open-frame structures, 133; partially-clad structures, 149; vertical vessels, 17, 155–158; wind load formulation, 7–8

heat exchangers. *see* air coolers
horizontal torsion, 93
horizontal vessels, 104–105, 159–161, 159*f*
Hurricane Celia (1970), 34, 35*f*
Hurricane Frances (2004), 38, 39*f*, 42*f*
Hurricane Harvey (2017), 33–34, 34*f*, 43*f*
Hurricane Hugo (1970), 34–36, 35*f*
Hurricane Humberto (2007), 36
Hurricane Katrina (2005): cladding damage, 44; cooling tower damage, 38, 41*f*–42*f*; tank damage, 36, 36*f*, 37*f*
Hurricane Rita (2005): cladding damage, 38*f*, 43*f*; cooling tower damage, 38; crane damage, 45*f*; electrical transmission structure damage, 38, 40*f*–41*f*; tank damage, 36; vertical equipment damage, 44, 44*f*
hurricanes: components and cladding performance, 38, 42*f*–43*f*, 44; construction load combinations and, 53–54; cooling tower performance, 38, 41*f*–42*f*; electrical power transmission performance, 36, 38, 39*f*–41*f*; existing structure considerations, 57; friction loads and, 62; live loads and, 62; operational gravity loads and, 57; performance of non-building structures, 33–44, 34*f*–45*f*; tall vertical equipment performance, 44, 44*f*, 45*f*; tank performance, 34–36, 35*f*–36*f*, 37*f*–38*f*
hydrostatic test loads, 54

impounding systems for LNG facilities, 68
industrial facilities. *see* non-building structures; *type of structure or facility*
International Building Code (IBC), 1, 49, 70

International Existing Building Code (IEBC), 70–71

lateral load on pipe bridges, 131–132
lattice Boltzmann method, 47–48
live loads, 62
LNG (liquefied natural gas) facilities: calculation procedures, 67–68; Code of Federal Regulations, 63–68; design guidance, 62–63; design method, 67; impounding systems, 68; regulation of, 62–63; wind loads, 62–68; wind speed and, 64–67
load combinations: alternate method for open-frame structures, 114, 115–117, 116*f*; ASCE 7-16, 49, 50*f*, 51*f*; for buildings, 49, 50*t*, 51*t*, 52; construction loads, 53–54; definitions, 52–54; empty loads, 54; existing guidelines, 49–52, 50*t*, 51*t*; existing structures, 57, 58*t*–59*t*, 60*t*–61*t*, 62; friction loads, 62; hurricane event considerations, 57; hydrostatic test loads, 54; IBC, 49, 50*f*, 51*f*; live loads, 62; new structures, 55*t*, 56*t*, 57; operational gravity loads, 57; PIP STC01015, 49, 50*t*, 51*t*, 52; supplemental guidelines, 52–62; upset loads, 54
longitudinal load on pipe bridges, 132–133

main wind force resisting system (MWFRS): design wind load calculation methods, 1–2, 77–78; open-frame structures and, 80–83; simplified procedure. *see* simplified procedure
Metal Building Systems Manual (MBMA 2010), 14
Minimum Design Loads and Associated Criteria for Buildings and Other Structures (ASCE 7-16), 1, 49, 64, 67
Model Code for Steel Chimneys, CCIND (2010), 27
modeling: computational fluid dynamics, 47–48; wind tunnel testing, 46–47
MWFRS. *see* main wind force resisting system

natural gas facilities. *see* LNG facilities
Navier-Stokes equations, 47–48
new structures: load combinations, 55*t*, 56*t*, 57; load definitions, 52–54; supplemental guidelines for, 52–62
non-building structures: additional information resources, 17–18; air coolers. *see* air coolers; background information, 5–20; components and cladding. *see* components and cladding; compressor shelters, 109; cooling towers. *see* cooling towers; cross-wind forces and, 10–11; development of design bases and procedures, 1–3; for electrical power transmission, 36, 38, 39*f*–41*f*; engineering concepts, 5–11; existing structures. *see* existing structures; flare towers, 44, 44*f*; formulation of wind load, 5–9; LNG facilities. *see* LNG facilities; new structures. *see* new structures; open-frame structures. *see* open-frame structures; partially-clad structures. *see* partially-clad structures; purpose of guidelines, 3; research priorities, 18–20; scope, 1; state of the practice, 3; vertical vessels. *see* vertical vessels; wind speed and, 6–7, 9–10

open-frame structures: aerodynamics, 11–12; along-wind force calculations, 134, 136–137, 138*t*–139*t*, 139–141, 140*t*, 141*t*; alternate method for wind load determination, 110–117, 112*f*, 118*f*,

145–146, 146*f*; application of wind load guidelines, 133–149; cross-wind force calculations, 141, 142*t*, 143, 143*t*–144*t*; design load cases, 90–93, 92*f*; force coefficients, 84–86, 86*f*–87*f*, 111–113, 113*f*, 115, 134–137; high-solidity structures, 117–120, 118*f*, 147, 147*t*–148*t*, 148*f*; horizontal torsion, 93; load combinations, 114–117, 116*f*, 120, 144, 144*f*, 146*f*; main wind force resisting system, 80–83; solidity ratio, 86–89; wind load determination, 80–93, 110–120, 134–147, 135*f*–137*f*, 144*t*
operational gravity loads, 57
ovalling of steel stacks, 28
overturning stability: construction load and, 53; empty load and, 54; operational gravity load and, 57; PIP STC01015, 48; roofs and, 26; tanks and, 25

partially-clad structures: aerodynamics, 13–14; application of wind load guidelines, 148–149, 150*f*; design wind load guidelines, 94–96, 95*f*
petrochemical facilities. *see* non-building structures; *type of structure or facility*
pipe bridges, 12, 129–133, 130*f*
Pipeline and Hazardous Materials Safety Administration (PHMSA), 62–63, 65
pipelines, 63, 65
pipe rack structures: application of guidelines, 124–129, 126*f*; cable trays, 127; design wind load guidelines, 78; open-frame design, 12; piping, 125, 127; structural members, 127–129, 128*f*; total base shear, 129
pipes and piping: design wind load guidelines, 78–80; large-diameter piping and vertical vessels, 104; pipe bridges, 131; pipe rack structures, 125, 127
PIP STC01015. *see* Process Industry Practices STC01015
platforms: horizontal vessels, 159–160, 161; vertical vessels, 152–155
pressure vessels: application of wind load guidelines, 149–161; design wind load guidelines, 96–106
Process Industry Practices STC01015: construction load combinations, 53–54; existing guidelines, 49, 50*t*–51*t*, 52, 69–70; existing structures and, 58*t*–61*t*, 62; new structures and, 55*t*–56*t*, 57

research priorities, 18–20
Reynolds number, 14–15, 48
rigid pressure models, 46
rigid vessels, 151–155

simplified procedure: for cooling towers, 29–30, 162–163; for vertical vessels, 97–101, 99*t*–100*t*, 151
solidity ratio, 86–89, 132
spherical vessels, 105–106
state of the practice, 3
steel stacks, 11, 27–28
storage tanks. *see* tanks
surface roughness, 6, 15, 72
sustained wind speed, 9, 64, 66

tanks: existing design practices, 22, 25–27; performance during hurricanes, 34–36, 35*f*, 36*f*, 37*f*–38*f*
temporary structure performance during hurricanes, 43*f*, 44
tower cranes, 44, 45*f*
turbulence, 48, 158

uncertainty in wind load analysis, 75
uplift forces, 25
upset loads, 54

velocity pressure: calculation of, 2; open-frame structures, 133, 134*t*; partially-clad structures, 149; tanks, 26; wind load formulation and, 5–6
vertical vessels: aerodynamics, 14–17; application of wind load guidelines, 150–158, 154*f*; appurtenances, 16; aspect ratio, 15–16; close spacing of, 103; cylinder aerodynamics, 14; design wind load guidelines, 96–104; detailed method for wind loads, 101–103, 151–155, 152*t*; directionality and, 7, 16–17; flexible vessels, 155–158; gust effects and, 17; large-diameter piping and, 104; neighboring vessels and, 16; performance during hurricanes, 44, 44*f*; Reynolds number, 14–15; rigid vessels, 151–155; simplified method for wind loads, 97–101, 99*t*–100*t*, 151, 151*t*; surface roughness, 15; vortex-induced vibrations, 17
vortex-induced vibrations, 17
vortex shedding: crosswind forces and, 11; pressure vessels and, 96; steel stacks and, 28

Welded Carbon Steel Tanks for Water Storage, AWWA D100 (2011), 22, 26–27
Welded Steel Tanks for Oil Storage, API 650 (2013), 22, 25
wind exposure categories, 71–72, 73*f*
wind loads: additional information resources, 17–18; air coolers. *see* air coolers; analysis uncertainty, 75; analytical determination of, 77–120; background information, 5–20; cable trays, 79, 80; calculation, 1–2, 77–78; CFD modeling, 47–48; components and cladding. *see* components and cladding; compressor shelters, 109; considerations and methods, 33–75; cooling towers. *see* cooling towers; crosswind forces, 10–11; development of, 1–3; electrical power transmission, 36, 38, 39*f*–41*f*; engineering concepts, 5–11; example site description, 123–124, 124*f*; existing practices, 21–30; existing structures. *see* existing structures; force coefficients. *see* force coefficients; formulation, 2, 2*e*, 5–9, 5*e*–6*e*; guideline updates, 77–109; horizontal vessels, 104–105, 159–161, 159*f*; hurricanes and. *see* hurricanes; LNG facilities, 62–68. *see also* LNG facilities; load combinations, 49–62. *see also* load combinations; new structures. *see* new structures; open-frame structures. *see* open-frame structures; partially-clad structures. *see* partially-clad structures; pipe racks. *see* pipe rack structures; piping. *see* pipes and piping; pressure vessels, 96–106, 149–161; purpose of guidelines, 3; research priorities, 18–20; scope, 1; spherical vessels, 105–106; state of the practice, 3; steel stacks, 11, 27–28; survey of existing practices, 21–22, 23*t*–25*t*; tanks. *see* tanks; vertical vessels. *see* vertical vessels; wind speed. *see* wind speed; wind tunnel procedure, 2, 44
Wind Loads and Anchor Bold Design for Petrochemical Facilities (1997), 21
Wind Loads for Petrochemical and Other Industrial Facilities (2011), 12
wind speed: interpretations of, 9–10; LNG facilities, 64–68, 66*f*, 67*f*; load formulation, 6–7; tanks and, 26
wind tunnel procedure, 2, 44–47